Quality Management System Handbook *for* Product Development Companies

Quality Management System Handbook *for* Product Development Companies

Vivek Nanda

CRC PRESS

Boca Raton London New York Washington, D.C.

Library of Congress Cataloging-in-Publication Data

Nanda, Vivek.
Quality management system handbook for product development companies / Vivek (Vic) Nanda.
p. cm.
Includes bibliographical references and index.
ISBN 1-57444-352-6
1. Quality control—Automation. 2. Manufacturing resource planning. 3. Production. planning—Data processing. I. Title.

TS156.N25 2005
658.4'013'011—dc22

2004058345

Direct all inquiries to CRC Press, 2000 N.W. Corporate Blvd., Boca Raton, Florida 33431.

Visit the CRC Press Web site at www.crcpress.com

International Standard Book Number 1-57444-352-6
Library of Congress Card Number 2004058345
Printed in the United States of America 1 2 3 4 5 6 7 8 9 0
Printed on acid-free paper

To my pride and joy — my son, Hersh Rahul Nanda.

PREFACE

You probably picked this book up because your company has recently launched or is considering launching a program to institute a quality management system (QMS). Or perhaps you are just curious about what it means to implement a quality management system in a company, what such an implementation entails, and what benefits it offers. This book describes a systematic approach for quality management and continuous improvement in a product development company by means of a formal quality management system.

The QMS implementation approach described in this book is centered on a high-level process for defining a QMS — from identification of essential prerequisites for a successful QMS implementation to mechanisms for continuous improvement. The QMS implementation process steps through the following five major phases:

- QMS implementation planning
- QMS definition
- QMS refinement
- QMS deployment
- Continuous improvement

The primary objective of this book is to provide pertinent information that a product development company would need to implement a QMS from scratch. Because this is intended to be a nuts-and-bolts book, it drills down deeper into each of the five phases. In addition to describing the high-level QMS implementation process and explaining *what* the elements of a sound QMS are, it also provides guidance on *how to* define, deploy, and continually improve a QMS. Implementation guidance is provided for all functional areas of a typical product development company, and without bias toward any particular industry or QMS standard. Several examples of QMS documentation are provided in the appendices to serve as ready reference to assist companies in defining their own QMS documentation.

Fundamentally, this book embraces the widely accepted view* that because much of the work in today's companies is accomplished by executing business processes, one must improve the business processes and supporting process management infrastructure in order to improve quality of products and services. Process management infrastructure includes but is not limited to organizational processes (including process documentation), standards, employee competencies, resources, and tools that support process management.

A process-oriented view of companies differs from the traditional function or line view of companies, wherein line managers are interested primarily in optimizing and improving their respective functions, as opposed to business processes (which typically span multiple functions). That is, improvements are driven vertically in each function of the company, as opposed to horizontally across the company. Exclusively focusing on optimization at a functional level generally leads to suboptimization at the organizational level, as each function focuses expressly on achieving its objectives. For example, if sales personnel focus solely on achieving their sales targets and make unrealistic commitments to customers to secure sales wins, then it will only result in customer dissatisfaction and long-term harm to the company's reputation.

It is important to recognize that product and service delivery are an outcome of interplay of cross-functional business processes. A process-oriented view of a company facilitates a better understanding of the various stakeholders involved in business processes and provides insight into how each function contributes to the quality of the final output. Such a view helps better manage the interfaces between various functions. Conversely, lack of process orientation causes the quality of products and services to fall victim to struggles between different functions that often operate with competing objectives. Such a situation is exacerbated if there are rigid personalities in key decision-making positions who ignore larger organizational objectives for the sake of personal ones. Personnel in such companies fail to see (or intentionally ignore) the big picture — the role various functions play in each process, and the value and contribution of each role in delivering a quality product or service.

For this reason, in recent years, the view has gained wide acceptance that the route to quality is via process management and process improvement. Companies have begun to realize that they must complement their

* It is acknowledged that Rummler and Brache were among the first advocates of such a view of organizations [RuB95].

functional view of business operations with a horizontal (or process) view of business operations. The restructuring and release of the most widely recognized QMS standard, ISO 9001:2000, founded on a process-based approach to quality improvement, is proof of the wider acceptance of the process view. Due to the increasing relevance of this process paradigm, this book takes a process-oriented approach to the establishment of a QMS, and describes quality practices that are relevant to typical processes in product development companies.

Finally, it is recognized that no two companies are alike. Companies differ with regards to:

- The industry they belong to
- Market segment (and product space) within the industry in which they operate
- Organizational structure
- Business processes they use
- Time to market (time from the conception of a product until its delivery to customers)
- Competency level of personnel
- Customer perception of quality
- Quality requirements placed on them as a result of customer or market demand

Therefore, the processes and related quality practices described in this book may not be applicable as is to all companies. One size does *not* fit all. Each company should examine the guidance provided in this book in the context of its operations and tailor it for its use, such that the established QMS is relevant to the company's operations. In addition, different companies will have different needs with regards to maturity and sophistication of their QMS and quality practices. For example, companies that develop mission-critical products or products that place human life at risk have the most stringent quality expectations and requirements. Therefore, when appropriate, companies should build upon the essential elements of a QMS presented in this book. Such enhancement should be as per organizational need and commensurate with applicable quality, regulatory, and customer requirements.

Vivek (Vic) Nanda, CSQE, CQA

THE AUTHOR

Vic Nanda has extensive experience in the software industry in various roles, working in quality assurance, system testing, and software development. To date, Vic has led the successful QMS implementation, registration, and maintenance at companies such as Ulticom, Architel Systems Corporation (a Nortel Networks company), and Ericsson. Vic's experience has included work in defining quality management systems as per ISO 9000:1994, ISO 9001:2000, TL 9000, and Capability Maturity Model (CMM) requirements. His interests in quality management systems include process definition, process architectures, corporate process assets, quantitative process improvement, project management, quality audits, supplier assessments, and customer satisfaction initiatives. He holds a Masters of Science degree (Computer Science) from McGill University, Montreal, and a Bachelor of Computer Engineering degree from the University of Pune, India. He is an ASQ Certified Quality Auditor (CQA), ASQ Certified Software Quality Engineer (CSQE), and a certified ISO 9000 Lead Auditor. Vic has authored several articles in quality and has presented at numerous international conferences. He also participates as a member of the review panel for various quality magazines and journals. **Vic can be reached at vic_nanda@hotmail.com.**

TABLE OF CONTENTS

PHASE III: REFINE

PHASE IV: DEPLOY

PHASE V: IMPROVE

APPENDICES

1

INTRODUCTION

This chapter provides an introduction to quality and thoughts of some of the pioneers in the field of quality. It explains what quality means in the context of a product development company,* and how quality can be made an integral part of a company's operations. That is, the deployment of quality practices in a company does not imply the implementation of a system that is on top of (or an adjunct to) the company's regular business processes. Quality cannot be an afterthought. It needs to be intrinsic to how a company performs and manages everyday operations. Implementation of quality practices entails organizational change because it requires that quality practices be *embedded* in the company's business processes. Implementation of quality practices in a company is the implementation of a more mature management system. Such a mature management system, built on a foundation of quality practices, is popularly referred to as a quality management system (QMS). An introduction to QMS is provided in this chapter. The chapter concludes with an overview of how this book is organized.

1.1 INTRODUCTION TO QUALITY

In technical usage, the word *quality* is widely accepted to have two meanings [Qua02]:

1. A characteristic of a product or service that bears on its ability to satisfy stated or implied needs; and
2. A product or service free of deficiencies.

* The word *company* and the word *organization* are used interchangeably in this book.

While there have been numerous books written on the subject, quality has remained a subjective term with several definitions. Even among the quality experts of the 20th century — P. B. Crosby, W. E. Deming, and J. M. Juran — there has been a lack of agreement on a definition of quality.

Broadly categorized, these quality experts' definitions of quality fall into two categories [HoH01]:

1. Quality is about satisfying applicable specifications. Quality is a simple matter of producing products or delivering services whose measurable characteristics satisfy a fixed set of specifications that usually are numerically defined.
2. Quality is about satisfying the customer. Independent of any of their measurable characteristics, quality products simply are those that satisfy customer expectations for their use or consumption.

Despite a lack of consensus amongst the quality experts on how to define quality, there are a number of similarities in their views on quality and on how to achieve quality improvements. Before we delve into quality and QMS, it is worthwhile to take a high-level view of quality and study what these quality pioneers have to say about it. This will facilitate a better understanding of what considerations and expectations are generally associated with a quality viewpoint. A clear understanding of what quality encompasses is necessary for one to identify what quality management and improvement activities should fall within the scope of a QMS, and what the extent of those activities should be. Again, it is important to recognize that the extent and maturity of quality activities appropriate for different companies *vary* based on factors described in the Preface of this book. It is for each organization* to determine the extent and sophistication of needed quality practices, based upon these factors, which are unique to every organization.

1.1.1 Philip B. Crosby: Four Absolutes of Quality Management and 14-Step Quality Improvement Plan

Crosby's views on quality fall in the first category. In his view, quality is conformance to requirements. For this purpose, he states that it is necessary to translate requirements into measurable product or service characteristics. With requirements stated in terms of numerical specifications, one can measure the characteristics of a product (e.g., diameter of a hole)

* In this book, organization refers to a company, enterprise, or institution (private or public) that provides a product to a customer. The term *organization* refers to "you" — the entity implementing the QMS.

or service (e.g., customer service response time) to see if it is high quality. In essence, this means that each requirement must be clear and unambiguously stated, and its outcome verifiable, so that it can be determined unequivocally whether the requirement has been satisfied.

One of Crosby's main contributions to quality was a set of four absolutes of quality management that provide insight into his quality philosophy. The following summary of his four absolutes is taken from the American Society for Quality [ASQ99]:

1. Quality has to be defined as conformance to requirements, not as goodness or elegance. Management must establish requirements, supply the wherewithal, and encourage and help employees to get the job done. The basis of this policy is "Do it right the first time."
2. The system for assuring quality is prevention, not appraisal. The first step to defect and error prevention is to understand the process* by which a product is produced. When a defect occurs, discovery and elimination are top priorities. Prevention is a knowledge issue for quality-focused workers.
3. The performance standard must be zero defects, not "that's good enough." The only performance standard that makes sense for "do it right the first time" is zero defects. Zero defects needs to be a performance standard for everyone in the company, from top management to workers on the line.
4. The measurement of quality is the price of nonconformance, not indices. A dollar figure can be established for the cost of quality (COQ) by determining the difference between the price of nonconformance and the price of conformance. The price of nonconformance is the expense of doing things the wrong way, and can account for 20 to 35% of revenues. Price of conformance is the cost of doing things right — typically 3 to 4%. Managers should spend time identifying where cost of quality is occurring and address what makes it occur.

Built on top of his four absolutes of quality management is Crosby's 14-step road map for quality improvement in an organization [Cro92]:

* Simply stated, a process is a sequence of activities performed that results in the transformation of inputs to outputs.

Management Commitment: Give clearly visible signals that management is committed to quality throughout the organization.
Quality Improvement Team: Set up quality improvement teams with senior representatives from each department to demonstrate high-level commitment.
Quality Measurement: Measure processes to determine where current and potential quality problems lie.
Cost of Quality: Evaluate the cost of quality, and explain its use as a management tool.
Quality Awareness: Raise the quality awareness and personal concern of every employee.
Corrective Action: Ensure a system is in place for analyzing defects in the system and applying simple cause-and-effect analysis to prevent recurrence.
Zero Defects Planning: Look for business activities to which zero defect logic should be applied.
Supervisor Training: Educate and train supervisors on their roles and responsibilities in the quality improvement process.
Zero Defects Day: Hold a Zero Defects Day to show everyone there has been a change, and to reaffirm management commitment.
Goal Setting: Encourage individuals to establish improvement goals for themselves and their groups.
Error Cause Removal: Encourage employees to communicate to management the obstacles they face in attaining their improvement goals.
Employee Recognition: Recognize and appreciate individual and team efforts for quality improvement.
Quality Councils: Establish Quality Councils to discuss quality matters on a regular basis.
Repeat the Cycle of Improvement: Do it all over again and emphasize that the quality improvement process never ends.

1.1.2 W. Edwards Deming: 14 Points

Deming is widely recognized as the guru of the Total Quality movement. His perspective on quality essentially falls in the second category — that quality should be viewed in terms of customer satisfaction [HoH01]. In Deming's view, a shift to a quality paradigm in a company is, in essence, a transition towards a more mature management style.

At the heart of Deming's quality philosophy is a set of 14 management principles (generally referred to as "Deming's Fourteen Points") that summarize his views on what a company must do to transition to a world-class company [Dem86]. These 14 points, without relevance to order of sequence, are as follows*:

1. Create constancy of purpose towards improvement of product and service, with the aim to become competitive, stay in business, and provide jobs.
2. Adopt the new philosophy. We are in a new economic age. Western management must awaken to the challenge, learn its responsibilities, and take leadership for change.
3. Cease dependence on inspection to achieve quality. Eliminate the need for inspection on a mass basis by creating quality into the product in the first place.
4. End the practice of awarding business on the basis of price tag. Instead, minimize total cost. Move toward a single supplier for any one item, on a long-term relationship of loyalty and trust.
5. Improve constantly and forever the system of production and service, to improve quality and productivity, and thus constantly decrease costs.
6. Institute training on the job.
7. Institute leadership. The aim of leadership should be to help people and technology work better.
8. Drive out fear so that everyone can work effectively for the company.
9. Break down barriers between departments so that people work as a team.
10. Eliminate slogans, exhortations, and targets for the work force. They create adversarial relationships.
11. Eliminate quotas and management by objectives (MBO). Substitute leadership. (Note: This is probably the most controversial of Deming's 14 Points. Deming contends that the problem with management by objectives is that such objectives push results at the expense of process. In order to get the necessary results on paper, managers will have incentive to distort information or distort the system. In the end, he argues, MBOs lead to deteriorating long-term results rather than improving them.)
12. Remove barriers that rob the hourly worker of his right to pride of workmanship.
13. Institute a vigorous program of education and self-improvement.
14. Put everybody in the company to work to accomplish the transformation. The transformation is everybody's job.

* Note that, over the years, Deming has modified the specific wording of various points, which explains the minor differences among the fourteen points as found in other publications.

1.1.3 J. M. Juran: Quality Trilogy and 10-Step Quality Improvement Process

Juran states that although the word quality has multiple meanings, two of those meanings dominate the use of the word [Jur88]:

1. Quality consists of those product features that meet the needs of customers and thereby provide product satisfaction.
2. Quality consists of freedom from deficiencies.

From this, it is evident that Juran's view of quality belongs to both the first and second categories. He accepts that although it would be convenient to have some definition for quality that is universally accepted, it is difficult to establish one. Therefore, he prefers to define quality by the short phrase "fitness for use." Here, use is apparently associated with customers' requirements, and fitness suggests conformance to measurable product characteristics.

One of Juran's most significant contributions to our understanding of quality is the Juran Trilogy®. It summarizes three primary managerial functions: quality planning, quality control, and quality improvement [GoD01].

Quality planning involves developing the products, systems, and processes needed to meet or exceed customer expectations. The following steps are required:

1. Determine who the customers are.
2. Identify customers' needs.
3. Develop products with features that respond to customer needs.
4. Develop systems and processes that allow the organization to produce these features.
5. Deploy the plans to operational levels.

Quality control involves the following key steps:

1. Assess actual quality performance.
2. Compare performance with goals.
3. Act on differences between performance and goals.

Quality improvement is an ongoing activity that never ends and involves the following key steps:

1. Develop the infrastructure necessary to make annual quality improvements.
2. Identify specific areas in need of improvement, and implement improvement projects.
3. Establish a project team with responsibility for completing each improvement project.
4. Provide teams with what they need to be able to diagnose problems, to determine root causes, develop solutions, and establish controls that will maintain gains made.

Juran recommends the following ten-step quality improvement process to achieve continuous quality improvements:

1. Build awareness of the need and opportunity for improvement.
2. Set goals for improvement.
3. Organize to meet the goals that have been set.
4. Provide training throughout the organization.
5. Carry out projects to solve problems.
6. Report progress.
7. Give recognition.
8. Communicate results.
9. Keep score.
10. Maintain momentum by building improvement into the company's regular systems.

In carefully reviewing the views of these quality experts, one can notice certain common threads:

Management responsibility: Management has a critical role to play in instituting quality in a company.

Quality awareness: Quality is everyone's job, and it is necessary to raise awareness of quality issues throughout the company.

Training: Training and education play a key role in developing a quality orientation in personnel, and in changing traditional mindsets.

Goals and measurements: Goals should be established for improvements, and progress should be measured and reported against the identified goals.*

* Note that on this issue, Deming differs from Juran and Crosby. He holds the controversial view that targets for the workforce should be eliminated because they lead to adversarial relationships, and that MBOs are ultimately detrimental to an organization.

Corrective and preventive action: Focus should be on doing it right the first time. Root cause analysis should be performed to effectively correct problems and to prevent new ones.

Continuous improvement: Quality improvements must be continual* and never-ending.

The aforementioned recommendations from the leading quality experts have become the cornerstones of today's QMS standards. Detailed discussion and implementation guidance on these essential elements of a QMS is deferred until later chapters.

1.2 QUALITY MANAGEMENT

In order to understand what it means to implement a QMS in an organization, one must first understand the meaning of the term *quality management.*

Quality management may be defined as follows:

Quality management comprises all activities that are required to plan for quality in an organization, and all activities that are required to satisfy quality objectives. Specifically, quality management comprises the following four elements (refer to Figure 1 for an illustration):

1. Quality planning
2. Quality control
3. Quality assurance
4. Quality improvement

Let us examine each of these four elements in detail:

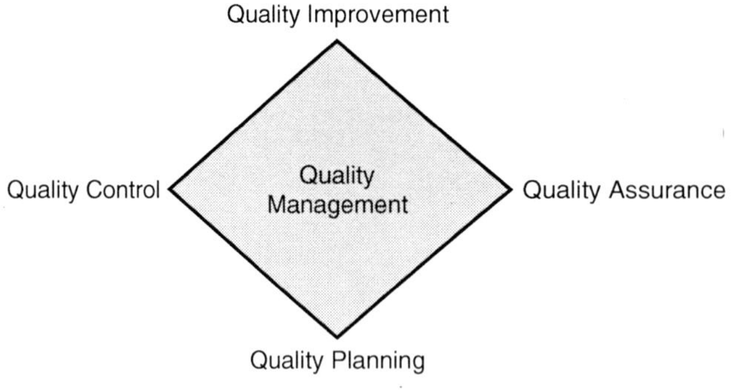

Figure 1 Quality Management

* In this book, the terms continual improvement and continuous improvement are used interchangeably.

1.2.1 Quality Planning

Quality planning refers to activities that are performed to:

Establish quality objectives — This includes establishment of both short-term and long-term quality improvement objectives (qualitative and quantitative). Establishment of long-term quality objectives demonstrates management vision and strategic thinking with regard to quality, while the establishment of short-term quality objectives facilitates prioritization of quality objectives in the near-term, and achievement of the long-term quality objectives.

Short-term and long-term are somewhat subjective terms, because they do not correspond to fixed implementation time frames. They instead depend on factors such as:

Product release cycle time of the organization — Release cycle time may be defined as the time period beginning with the initial identification of product requirements and ending with delivery of finished product to the customer. In other words, it is the duration of a typical product development project in a company. For example, say the release cycle time of Company A is 6 months and that of Company B is 12 months. If each company establishes a quality objective that is applicable to one project (short-term quality objective), then the short-term quality objective is applicable over a period of 6 months in the case of Company A, while it is applicable over a period of 12 months in the case of Company B.

Feasibility of achieving a particular quality objective within a specific time frame — Often, a desired quality objective may be too difficult or stringent to achieve in the near-term (keeping in mind the current state of the practice or maturity of the organization). In such a case, the quality objective may be set as a longer-term objective that is to be met over a relatively longer time period (where the time period is determined by the organization), and its achievement may be planned incrementally so that there is continual improvement toward the established objective. This can be accomplished by decomposing the long-term quality objective into several smaller feasible quality objectives that can be met in successive short-term increments (where the duration of each short-term increment is determined by the organization).

A short-term quality objective is one that is applicable to one product development project, or applicable over a period of a few months. For example, say a company's historical data reveal an average of 1000 defective products per million. A feasible short-term quality objective may be a 10% reduction in defective products, achieved by the following year.

For the same product, a longer-term quality objective would be one to be met over a longer time period, say three to five years. For example, the organization may establish a long-term quality objective to achieve, over a five-year period, a 50% reduction in defective products (see Figure 2). Similarly an organization could identify the short-term objective of increasing its overall customer satisfaction rating (as measured from customer satisfaction surveys) from the current score of 75% to 80% by the end of the calendar year, with the long-term objective of gradually improving customer satisfaction ratings to 90% over a three-year period.

Identify quality requirements — For a product development company, there are essentially two types of quality requirements:

Quality requirements applicable to processes — these we can refer to as "process quality requirements" (also commonly referred to as "QMS requirements"); and

Quality requirements applicable to products — these we can refer to as "product quality requirements."

Quality planning covers the identification (or establishment) of both these types of quality requirements.

Because QMS requirements are applicable to an organization's business processes, such quality requirements demand compliance during process

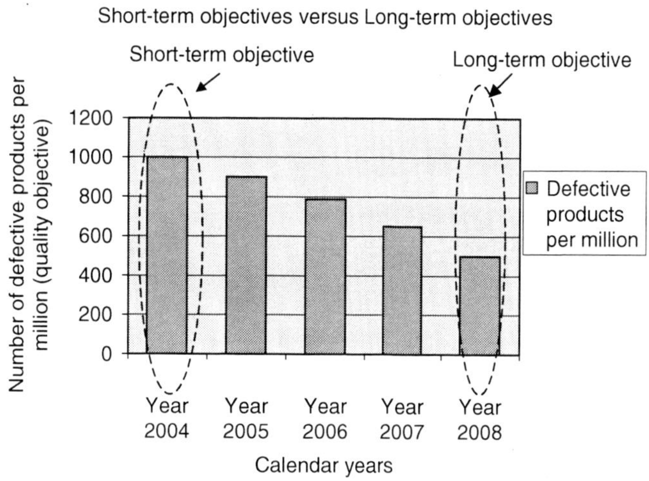

Figure 2 Short-term vs. long-term quality objectives

execution. There are QMS standards* that contain standardized QMS requirements, such as the widely recognized ISO 9001:2000 standard [ISO9001]. An organization might identify a QMS standard to which it will adhere, and/or choose to specify QMS requirements internally. Even organizations that select a quality management system standard for use still need to define certain internal QMS requirements as per their unique needs, customer requirements, business processes, organizational structure, and upper-management direction. QMS requirements defined internally by an organization are specified as *shall* or *should* requirement statements** in the relevant QMS documentation associated with the respective business process. Verification of adherence to QMS requirements is performed by means of quality audits, monitoring of processes in real time, and examination of process outputs.

Product quality requirements apply to a specific product (or family of products), or to a specific release of the product. Product quality requirements may be quantitative or qualitative. For example, a product may have "qualitative" quality requirements related to user-friendliness and "quantitative" quality requirements related to reliability of the product. If a product is built for the mass market, the organization internally specifies product quality requirements by considering market needs, and describes them in a formal requirements document. On the other hand, if a product is built to contract, as per the needs of a specific customer, the customer and the supplier agree on the functional and quality requirements for the product that are formally documented in a contractual agreement or requirements document. In both scenarios, the product is validated after development by testing or inspection against applicable requirements to verify conformance. When required, appropriate corrective action is initiated.

Plan for a Quality Management System — For a product development company, planning for a QMS entails planning for all elements of the QMS that are necessary to meet requirements for quality. This includes but is not limited to planning for:

* Organizations can be formally registered to a QMS standard by complying with all requirements in the standard, and after being successfully assessed by an independent accredited registrar.

** Note that all *shall* and *should* statements in QMS documents are not necessarily internally specified QMS requirements. Several such requirements may have been propagated from the applicable QMS standard. Regardless of whether a requirement originates from a QMS standard or is internally specified, all *shall* and *should* requirements in the applicable QMS standard *and* in the organization's QMS documentation require compliance during process execution.

- Establishment of product development and support processes
- Establishment of control points (generally in the form of milestones) and entry and exit criteria associated with intermediate stages
- Definition of methods
- Establishment of workmanship standards
- Identification of required resources
- Identification of work products (including intermediate work products)
- Establishment of guidelines for tailoring of product development processes; for instance, within the context of a specific product development project

Plan for process execution (in accordance with the QMS) — This entails planning for the *application* of the defined QMS. This means planning for process execution in accordance with the defined QMS such that requirements for quality can be met. Typically, such planning for process execution is performed in the context of planning for the development of a product or execution of a contract. The output of such planning is not one plan document, but generally a set of planning documents, such as project plan, product development plan, product test plan, and so on. For example, preparation for product testing generally entails preparation of a product test plan to test the product in accordance with the organization's product test process (defined as part of its QMS).

1.2.2 Quality Control

Quality control comprises activities executed to fulfill requirements for quality. This includes:

Activities to monitor a process to ensure its output is of required quality; and
Activities to correct discrepancies when they occur.

Figure 3 illustrates the process flow associated with a typical quality control activity:

A process (or activity) is executed and its output verified to ensure it is of desired quality.

The verification of process output is performed by comparing the process output with applicable specifications, standards, or requirements.

Discrepancies (or anomalies) observed in the output are regarded as deficiencies that need to be corrected.

Observed discrepancies are corrected by determining and implementing suitable corrective actions and, when appropriate, preventive actions (to prevent similar discrepancies in the future). Suitable corrective action

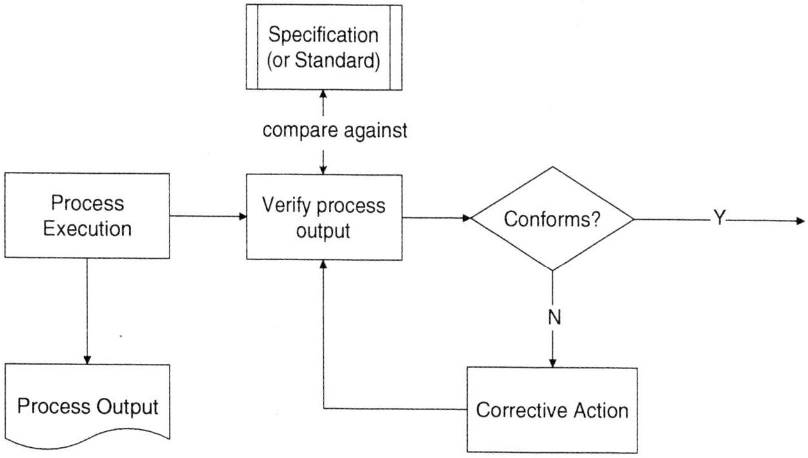

Figure 3 Quality control

may entail rejecting or reworking the process output, or correcting the process (or activities) that resulted in the deficient output.

The corrected process output is reexamined to verify that the previously observed deficiency no longer exists. In certain circumstances, it may not be possible to take corrective action; for example, when the defect is marginal, corrective action is not cost-effective, or there are other reasons why corrective action cannot be taken. In such circumstances, it generally is sufficient to perform risk mitigation for requirements not met, if the related risks are significantly low, or otherwise acceptable to the customer.

Quality control is not necessarily performed only on the resulting process output at the end of a process. When appropriate, it should also be performed *during* process execution to facilitate timely corrective and preventive action.

Generally, quality control activities are regarded as reactive because their primary purpose is to detect and eliminate defects already in a product, as opposed to quality assurance activities, which generally are regarded as proactive because they primarily seek to prevent defects in a product.

1.2.3 Quality Assurance

Quality assurance may be defined using the following widely accepted definition:

Quality assurance comprises all the planned and systematic activities implemented within the quality system that can be demonstrated to provide confidence that a product or service will fulfill requirements for quality [Qua02].

There are two primary parties that have a need for quality assurance: management and customers. Both these parties have a vested interest in determining beforehand that requirements for quality will be met. Quality assurance, by definition, cannot guarantee that requirements for quality will be met. However, the extent of quality assurance has a direct bearing on the measure of confidence that management and customers have that quality requirements will be met. Quality assurance can help provide such confidence to management and customers by demonstrating that:

- Plans exist for the achievement of quality requirements; e.g., a Project Quality Assurance Plan provides information on the planned quality assurance and quality control tasks in a specific product development project.
- Means exist that specify how requirements for quality are to be achieved, such as procedures (also called "standard operating procedures," or SOPs) and work instructions (including identification of required resources, such as employee* skills and equipment), methods, checklists, etc.
- Means necessary to achieve quality requirements are available and well deployed in the organization. For example, quality audits (internal or external) serve as a useful mechanism to verify whether or not defined processes are being adhered to, to verify competency of personnel performing the work, and to verify whether or not resources required to perform the work are available.
- Means to achieve quality requirements have been demonstrated to be adequate for meeting requirements for quality. This can be accomplished by reviewing quality records and process and product measurements data and trends.
- Means exist to correct discrepancies when they occur (quality control activities); e.g., product reviews (or inspections) and product testing.
- Means exist to assess quality-related risks continually, and to plan for risk mitigation and risk contingency — for example, project review meetings and milestone review meetings.
- Means exist to verify that all quality requirements have been met prior to release of the product or service to the customer, and that necessary corrective action will be taken when such requirements are not met (as explained previously under quality control).

* Throughout this book, the word employee is used to refer to any individual working for the organization, regardless of whether the person works full-time or part-time, and regardless of whether the person is permanent, temporary, or on contract. The word employee is use interchangeably with personnel.

Finally, it is worth pointing out that the terms quality control and quality assurance quite often are used interchangeably to refer to actions performed to ensure the quality of a product or service. This is because some people argue in favor of a broader definition of quality assurance, that it not only provides assurance that requirements for quality will be met, but also provides assurance that requirements for quality have been met. If one adopts this broader definition of quality assurance, it is obvious that certain quality control activities also help provide the assurance that quality requirements are being and have been met. For example, results of successful testing or inspection of a product against product requirements (examples of quality control activities) provide management the assurance that requirements for quality have been met. Attempts to accurately classify each activity performed to ensure the delivery of quality product or service, as either a quality control activity or a quality assurance activity, is perhaps a pointless exercise. It is preferable to recognize that the concepts of quality assurance and quality control collectively encompass the following:

- Achievement of quality requirements has to be planned;
- Adequate means to achieve quality requirements have to be provided and deployed throughout the organization; and
- Achievement of quality requirements has to be monitored continually (and appropriate action taken, when necessary).

As long as an organization recognizes the above underpinnings of a sound QMS, it is well on its way to establishing one.

1.2.4 Quality Improvement

Before we examine what it means to achieve quality improvement, it is necessary that the terms *efficiency* and *effectiveness* be understood. Both these terms are widely used in most definitions of quality improvement. Efficiency refers to savings in time, money, and effort expended to accomplish a task. Effectiveness refers to the goodness or quality of an accomplished task. Efficiency and effectiveness generally are regarded as being inherently incompatible because improvements in efficiency usually imply sacrificing effectiveness, and vice versa. However, elegant solutions to quality problems are those that result in a quality improvement by achieving a balance between efficiency and effectiveness.

Quality improvement may be defined as:

- Enhancement in the effectiveness and efficiency of processes; and
- Enhancement in the extent to which a product satisfies applicable requirements (including quality requirements).

Both enhance an organization's ability to meet customer expectations, and thus translate into improved customer satisfaction.

A customer may be internal (within the organization) or external. There might be quality improvements that result in an increase in operational efficiency and effectiveness within the organization, or result in the reduction of defects in an internal work product, both of which might not be directly visible to an external customer. Generally, however, internal quality improvements externally manifest themselves. For example, an organizational objective to resolve product defects *efficiently* (here, the consideration is long-term efficiency) would generally result in the dedication of more time to perform additional testing of the product (and thus find more product defects prior to delivery to the customer). This would result in savings in time, money, and effort due to the need to fix fewer product defects after release.

Similarly, an organizational objective to resolve product defects *effectively* might necessitate a classification of each product defect into an appropriate category. This might be followed by an investigation to identify which types of product defects are the most prevalent (or severe), and how each type of defect can be minimized as a whole (beginning with the type of defect that is the most prevalent and/or most severe, as appropriate). Such an analysis, referred to as a Pareto analysis, goes beyond the short-sighted approach of merely fixing each reported defect, without an analysis of why that defect occurred in the first place and how prevalent are similar types of defects.

With such a tactical view to resolving defects, an opportunity to effect process change (here, process improvement) that might have helped prevent or minimize similar defects in the future would be lost. Note that as in the previous example, the customer would realize the impact of such a quality improvement effort due to a gradual decline in the number and severity of defects in the finished product.

One should bear in mind that a quality improvement, though effective and efficient in the long-term, might reduce operational efficiency in the short-term. This is because the cost of an improvement effort is immediate, while the corresponding benefits generally are delayed. Therefore, any decrease in near-term operational efficiency, which is more readily apparent, could, due to lack of foresight, result in reluctance and resistance to adopt the change. Unfortunately, this challenge is a natural accompaniment to any quality improvement effort.

The Deming cycle (also referred to as the Plan Do Check Act (PDCA) cycle [Dem86]) provides a high-level framework that can be followed for defining an effective quality improvement process. The PDCA cycle consists of the following four phases:

- Plan (a change or improvement)
- Do (implement the change or improvement)
- Check (verify that it meets desired results)
- Act (deploy the change, modify and reverify the change, or abandon it)

A quality improvement process that is derived from the PDCA cycle essentially consists of the following steps:

1. Define a quality improvement goal for the process or product to be improved. Ensure that the identified goal meets the SMART* criteria (Specific, Measurable, Acceptable, Realistic, Timely), which are explained in Chapter 2.
2. Plan
   a. Explore possible solutions that will effectively and efficiently achieve the stated goal.
   b. Prepare an improvement plan that specifies:
      i. How the improvement goal will be achieved — that is, which solution has been chosen for implementation.
      ii. A list of planned actions and personnel responsible.
      iii. An implementation time frame.
      iv. Progress review and reporting mechanisms.
   c. Plan for anticipated resistance to change, and strategies to overcome it.
   d. Secure all resources necessary to execute the plan.
3. Do
   Execute the improvement plan. When appropriate, first implement the improvements as a pilot project, as opposed to directly implementing a permanent process change.
4. Check
   Review the results of the improvements against expected results, and when actual results differ from expected results, plan for appropriate corrective action.
5. Act
   a. If necessary, implement the corrective actions that are necessary to achieve expected results.

* Adapted from S.M.A.R.T (Specific, Measurable, Attainable, Realistic, Tangible) goals from Paul J. Meyer's "Attitude is Everything" (http://www.topachievement.com/smart.html).

b. Determine whether or not the improvement goal has been achieved. If the improvement project was successful, establish controls that are necessary to hold the gains achieved and to perform at the improved level of performance consistently in the future.
c. If the improvement project was conducted as a pilot, determine future action based on pilot results:
   i. A successful pilot will typically result in the deployment of the piloted solution (by effecting a permanent process change, where needed).
   ii. An unsuccessful pilot (where the improvement goal was not achieved) will typically result in:
   iii. Revisions to the piloted solution, followed by additional piloting activity;
   iv. The complete abandonment of the piloted solution in favor of an alternate solution that is then piloted; or
   v. The abandonment of the improvement goal, or its deferral to the future.

Finally, it is worth noting that, somewhat recently, quality improvement and process improvement* have come to be widely acknowledged as synonymous. This is due to the growing realization that companies develop products and provide services by executing business processes. Therefore, improvements in the quality of products and services result from improvements in product and service realization processes. Consequently, over the recent years, focus has shifted towards defining, measuring, analyzing, monitoring, controlling, and improving business processes, in order to achieve and maintain improvements in product and service quality. It is also accepted that the quest for quality improvement is never-ending, and best-in-class organizations are those that strive for and demonstrate *continual* quality improvement.

1.3 QUALITY MANAGEMENT SYSTEM

1.3.1 What Is a QMS?

A *quality management system* (QMS) is the means by which quality management practices, such as those described in the previous section, are made an integral part of an organization. A QMS is not a temporary fad, but a permanent part of an organization with a direct bearing on how the organization conducts its business. QMS is not a vague phrase;

* *Process improvement* refers to improvements in process capability and performance.

it has a very specific meaning: a QMS has a structure, a defined scope, responsibilities, necessary content (in terms of defined processes and supporting QMS documentation), and required resources to accomplish quality planning, quality control, quality assurance, and continuous quality improvement activities. If an organization merely implements a few quality management practices in its operations, it cannot claim to have a quality management *system* in place. A QMS is not static, and by definition it must be improved continually in order to enhance organizational effectiveness and efficiency. It may be formally defined as follows:

> A quality (management) system consists of the organizational structure, procedures, processes, and resources needed to implement quality management [ISO8402].

1.3.2 Reasons for Implementing a QMS

For most organizations, the primary motivation for implementing a QMS is either management need or customer demand. Management's motivation for implementing a QMS usually stems from its need to improve productivity, improve product quality, and reduce time-to-market, thus gaining a competitive advantage. Sometimes, management's motivation for implementing a QMS is driven by competitive pressure, where the organization's competitors have established (or are in the process of establishing) a formal QMS with the goal of registration to a recognized QMS standard, such as ISO 9000. In such cases, registration to a quality management system standard is perceived to be a valuable asset for marketing and soliciting new customers.

Customer demands on an existing supplier (or a potential supplier) to implement a QMS is driven by the customer's need of an assurance that the supplier is capable of meeting the customer's quality requirements. Often, such a demand may be made in response to continued subpar performance of an existing supplier, or prior to approving a new supplier. In certain industries, customers (including government agencies) also go to the extent of inviting bids only from suppliers who have attained a particular quality registration. For example, many buyers in the European Union (EU) expressly require their suppliers to be ISO 9000–registered. The U.K. Ministry of Defence, the U.S. Department of Defense, the National Aeronautics and Space Administration (NASA), and the U.S. Food and Drug Administration (FDA) require their large contractors to be ISO 9000–registered. In addition, automotive companies such as Ford, Chrysler, and General Motors impose the industry-specific QMS standard, QS-9000, on their suppliers [Rom00]. Because an organization that does not have a QMS in place may be barred from bidding for potential business, it is likely to translate into management motivation to implement a QMS.

It has been argued that the "management-motivated approach will normally be more comprehensive and fruitful than the model used for demonstrating the adequacy of the quality system" (i.e., the customer-motivated implementation of a QMS) [ISO9000-1]. In other words, the likelihood that a QMS will be adequate and effective is significantly improved if its implementation is driven by internal motivation in the organization (management need) rather than external pressures (customer demand). In fact, as is explained later, in Chapter 2, management commitment to quality is the most significant prerequisite for a successful QMS implementation. When management visibly demonstrates its commitment to quality, and promotes a quality-oriented and customer-focused mind-set in the organization, it encourages the employees to strive to realize the true benefits of the implemented QMS. On the other hand, a QMS that is implemented solely with the objective of achieving a coveted quality registration to win new business, or please potential customers, will serve merely as a short-lived marketing tool. This is because the lack of an *effective* QMS eventually will manifest itself in poor product quality, late product delivery, low employee morale, and dissatisfied customers.

1.3.3 Benefits of Implementing a QMS

Implementation of a QMS in an organization offers near-term and long-term rewards:*

1. Defined processes and supporting QMS documentation are the basis for repetition, and help reduce (and eliminate) variation within process execution. As variation is reduced, it results in improvements in operational efficiency.
2. With the implementation of corrective and preventive solutions that effectively address the root causes of quality problems, permanent solutions are implemented. This results in improvements in organizational effectiveness.
3. A QMS enables an organization to focus on how it executes its business processes. Such process focus and awareness are essential in order to be able to monitor and analyze process performance for continual improvement.
4. A QMS fosters continual improvement in the organization's productivity, rework costs, on-time delivery performance, and within budget project execution. This enables the organization to enhance its bottom-line revenue growth.

* Some of the listed benefits are reproduced with permission from [Nan03].

5. A QMS results in higher-quality products and services, as quality management practices are continually improved.
6. As an organization improves the quality of products and services, it improves customer satisfaction levels, which helps improve customer loyalty and customer retention.
7. A QMS enables the organization to gain a competitive advantage due to its being perceived as a "best-in-class" supplier by its customers. This enables the organization to retain customers, attract new ones, increase market share, and enhance top-line revenue growth.
8. A QMS enhances an organization's competitive position by allowing it to present itself as a viable supplier in situations where a customer requires its suppliers to have a formal QMS in place (although in certain cases customers also seek registration to a QMS standard, as discussed previously).
9. A QMS enhances customer confidence in the ability of a supplier to deliver products and services according to specified quality requirements (quality assurance).
10. A QMS reduces the organization's reliance on "heroes" to make projects a success, because all employees are aware of the required quality management practices. In other words, it enhances an organization's ability to achieve quality requirements because employee competencies are augmented by a process infrastructure that helps achieve the identified requirements.
11. A QMS reduces (or eliminates) an organization's dependence on a few individuals for information regarding critical processes, because such processes are now formally documented. This reduces organizational vulnerability to employee turnover.
12. A QMS reduces waste of resources and loss of reputation resulting from rejection and rework of inferior-quality products (referred to as *Cost of Poor Quality*). This enables the organization to shift from a reactive mode of operation (performing corrective action) to a proactive mode (performing preventive action).
13. A QMS promotes employee understanding that quality is everyone's responsibility. The realization that each employee contributes to the achievement of quality requirements helps institutionalize quality improvements across the organization, at all levels.
14. Employee morale and satisfaction improve as employees participate in defining their processes, and are empowered to own, monitor, and continually improve those processes.
15. A QMS results in improved communication both internally and externally, which results in improvements in efficiency and effectiveness, and improved customer–supplier relations.

The aforementioned benefits are depicted in Figure 4 — both from an internal perspective (that is, from within the organization) and an external perspective (that is, from a customer viewpoint). Note that Figure 4 is intended only to show significant direct and derived benefits of implementing a QMS, and to classify them broadly under the perspective (internal or external) from which they are realized. This is not intended to be an accurate classification — that is, certain benefits that are shown in the internal perspective may be realized from an external perspective as well, and vice versa. Nevertheless, the figure shows that from an internal organizational perspective, key benefits realized from implementing a QMS include process-driven improvements in organizational effectiveness, reduced process variation, and greater process predictability. Streamlined and improved processes minimize rework and result in improvements in productivity, quicker time-to-market, and cost savings from time, labor, and material saved. In the end, this translates to bottom-line improvements for the organization.

From an external perspective, improvement in product and service quality, in on-time delivery performance (due to mature, stable, and repeatable processes), and in organizational effectiveness result in improved customer satisfaction levels (for existing customers). A satisfied customer is more willing to provide positive references for its supplier, and is more likely to repurchase from the supplier. Therefore, a satisfied customer is less likely* to defect from the supplier. Customer defection rate is a key measure of the commitment of the customer to its supplier, better known as customer loyalty. A loyal customer offers long-term rewards in the form of future repurchase and, generally, steady and incrementally improving purchase volumes. In addition, an organization that is renowned for the quality of its products and services establishes itself as a best-in-class organization. Such a best-in-class reputation enhances the organization's image, which is also dependent on other factors, such as the organization's market leadership and innovativeness. An organization's image often becomes the key differentiating factor during purchase decision making by potential customers. In addition, positive references from customers and the organization's demonstrated ability to retain customers, including marque customers, enable the organization to win new customers. In the end, repeat business from existing customers and business from new customers contribute to gradually increasing top-line revenue growth for the organization.

* Assuming the supplier's product offers capabilities similar to those of its competitor's product and the product is competitively priced.

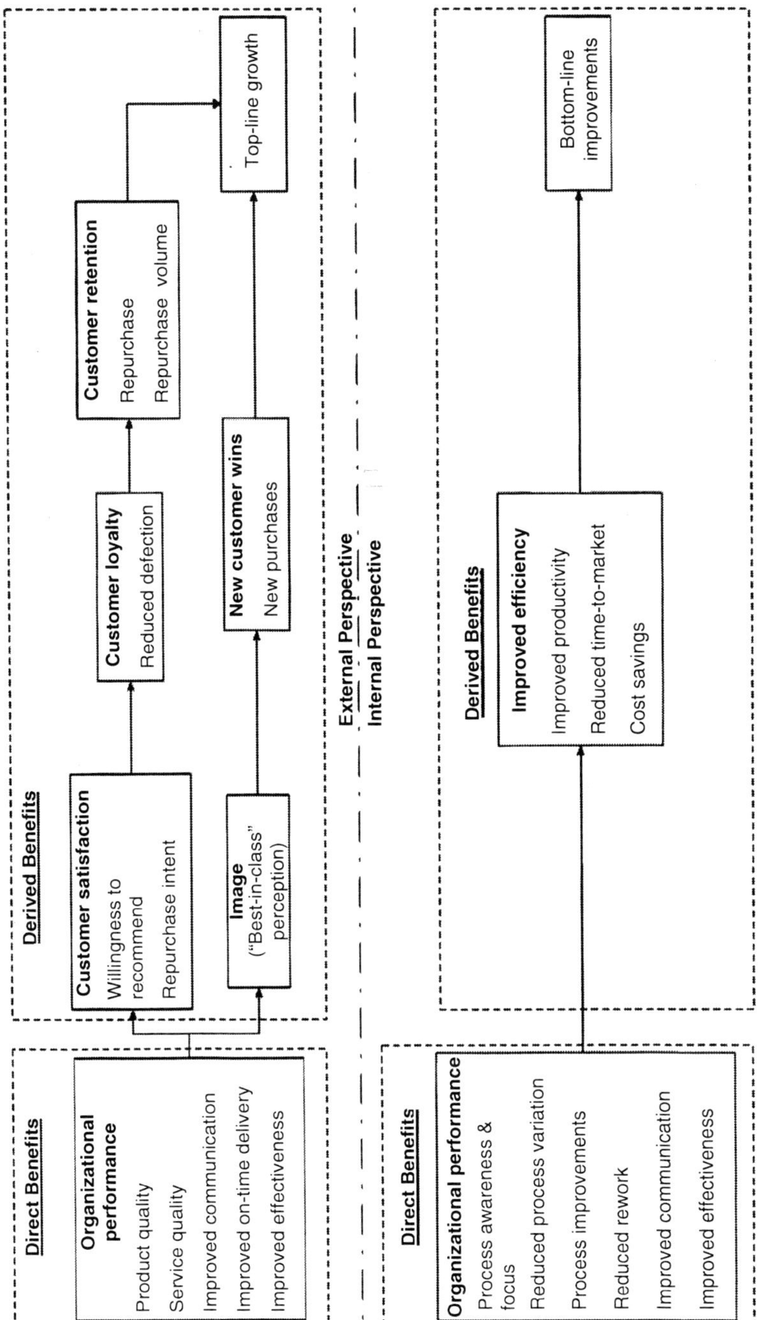

Figure 4 Benefits of Implementing a Quality Management System

1.4 ABOUT THIS BOOK

This book is structured as a "process flow"; that is, the chapters step through the logical sequence of a recommended five-phase QMS implementation process. Following is a high-level summary of the chapters:

Chapters 2 and 3 describe the first phase of the QMS implementation, which deals with *planning* for a successful implementation. Chapter 2 provides guidance on overall implementation strategy, implementation process, and mechanisms for monitoring, controlling, and reporting progress. Chapter 3 describes a critical element of undertaking a QMS implementation — planning for QMS documentation. Different types of QMS documentation are explained, along with document management and control mechanisms.

Chapters 4 and 5 describe the second phase of the QMS implementation, which deals with *defining* the QMS. Chapter 4 explains such fundamental concepts as process mapping, creation of process documentation, and means for tailoring standardized organizational processes for specific use (for example, for use in a particular project). Chapter 5 goes a step further and introduces widely used quality practices pertaining to different functional areas of product development companies. Note that quality practices and processes that pertain to functional areas such as finance, sales, marketing, strategic alliances, legal issues, human resources,* and other ancillary functional areas are beyond the scope of QMS standards for product development and support, and therefore beyond the scope of this book. However, companies can define the processes and needed quality practices for such functions using relevant concepts from various chapters, as appropriate. For example, a company's marketing department can define its operational processes by applying the process mapping and related concepts explained in Chapter 4. Similarly, it can identify suitable process metrics for the defined processes by following the related guidance on process measurements in Chapter 8.

Chapter 6 explains the third phase of the QMS implementation process — ensuring the internal consistency, correctness, and completeness of the defined QMS, and *refining* it, as needed, to ensure that the aforementioned criteria are met. Note that it is not recommended that such an activity be only performed once the QMS has been completely defined. Such an activity must be an inherent part of the second phase, in order to ensure that defined processes and quality practices are coherent and consistent. However, a dedicated phase to perform *final* verification and validation is necessary because, typically, organizational process definition** and QMS development proceed simultaneously throughout the various

---

* Except for activities related to employee competency assessment and development that are described in this book.

functional areas of the organization. This increases the likelihood of a *disconnect* (or inconsistency) between related processes in different functional areas. Or, those defining the processes in one functional area may be oblivious to the interdependency (or impact) with another functional area. So process definition in one functional area may unknowingly threaten assumptions or expectations in another functional area, and have an adverse impact on interdepartmental process interfaces. In other words, while defining *individual pieces* of the QMS, and ensuring that they are correct, complete, and consistent, it must be ensured that *all the pieces* (once developed) *taken together as a whole* are complete, correct, and consistent. Ensuring the latter is the purpose of Phase III of the implementation process.

Chapter 7 explains the fourth phase of the QMS implementation phase, which deals with *deployment* (or institutionalization) of the defined QMS. Various means of promoting and verifying deployment of a newly established QMS are described in this chapter. While such an activity may be performed in a dedicated phase, organizations may also perform a phased rollout, wherein pieces of the QMS are incrementally defined and deployed in the organization. When a phased rollout is performed, it has implications on how to refine the complete QMS, as certain pieces of the QMS might already have been deployed while other pieces were in the process of being defined and/or deployed. Implications of adopting such a phased rollout on the refinement of the QMS are discussed in Chapter 6.

Chapter 8 describes the final never-ending logical phase of the implementation process that pertains to continuous improvement of the QMS. Various means of achieving continuous improvement gains are described, along with how all the key elements of the QMS fit together to form the cycle of continuous improvement. Means of deploying the gains achieved via continuous improvement are also described.

The book concludes with an appendix section that provides several examples of QMS documentation, including a sample Quality Manual outline, examples of procedures, forms, templates, audit checklist, and examples of QMS training material for employee training.

** The term *process definition* refers to the prescribed way of executing a process which is not necessarily documented. Due to the cost associated with creating and maintaining process documentation, it is not feasible for an organization to document all its processes. For processes that are undocumented, the process owner or relevant management-level person's description of a process constitutes the process definition to which process practitioners are expected to conform.

PHASE I

PLAN

2

QMS IMPLEMENTATION PLANNING

One of the critical first steps and an essential prerequisite for a successful QMS implementation is detailed implementation planning. QMS implementation planning includes activities such as:

1. Identification and satisfaction of prerequisites for a successful implementation
2. Definition of an implementation strategy and process
3. Preparation of a formal implementation plan
4. Establishment of mechanisms to monitor, control, and report implementation progress

An explanation of these topics is needed for several reasons. If an organization embarks upon a QMS implementation with a well-thought-out execution plan, it significantly simplifies the task of implementing a QMS, because the organization is able to tackle the goal of implementing a QMS in a piecemeal manner. Detailed implementation planning facilitates the decomposition of the final objective into smaller and achievable intermediate objectives. Consequently, implementation activities are purposeful and manageable during each phase of the implementation. Conversely, lack of adequate implementation planning likely will result in problems, including but not limited to:

- Lack of foresight regarding critical success factors for the QMS implementation and, subsequently, organizational inability to ensure prerequisites are in place to maximize chances for success
- Poor employee awareness of the roadmap toward the identified goal, resulting in a lack of a sense of direction or ability to establish implementation priorities

- An inability to ensure that needed resources have been secured
- Inadequate, excessive, inconsistent, or inaccurate QMS documentation due to lack of a well thought-out implementation strategy (which has a direct bearing on what QMS documentation is produced, how much of it is produced, and how it is produced)
- Repeated rework to address deficiencies in QMS implementation due to inadequate forethought
- Inability to assess adequacy of progress (against a plan)
- Schedule delays due to poor visibility into what detailed tasks need to be executed and how much effort each task entails
- Deficient deployment and inadequate employee awareness of the defined QMS
- Problems associated with employee morale arising from lack of awareness of implementation strategy, process, and plans. (These in turn may negatively impact employee participation during QMS implementation and thus impede progress, or they may discourage employee use of the established QMS.)

In order to preclude these problems, an organization must invest a significant amount of effort in planning its QMS implementation. The output of this exercise then serves as the blueprint for the subsequent QMS implementation. Due to the significance of QMS implementation planning, this activity is the *de facto* first phase in any QMS implementation from scratch, and so it is in this book as well. A critical piece of QMS implementation planning is planning for QMS documentation, which due to its significance and extent is discussed separately in the next chapter. This chapter focuses on all other QMS implementation planning aspects.

2.1 IMPLEMENTATION PREREQUISITES

In order to understand what prerequisites need to be satisfied for a successful QMS implementation (in addition to the prerequisite of implementation planning), it is important to understand how success in this context may be defined. Simply stated, QMS implementation is successful if the defined QMS is adequate for the organization's needs, well deployed (adhered to) in the organization, effective, and continually improved*.

* Initially, however, when the QMS is newly established, emphasis is on establishing mechanisms for continual improvement. The objective of continual improvement itself is a longer-term objective. Therefore, initially, success may be gauged by whether or not mechanisms are in place to facilitate continual improvement, with perhaps some early examples.

Working backwards, one can identify the critical success factors — that is, prerequisites for success. For example, if the QMS is to be adequate for the organization's needs, a critical success factor is that QMS processes be defined with input and review feedback from relevant employees. Here, relevant employees are those who execute a process and those who interface with the process. "Employee involvement" (or participation) therefore becomes an essential prerequisite for a successful QMS implementation. Using the aforementioned approach, organizations may perform an exercise during implementation planning to identify prerequisites for success as per their unique needs. Generally, however, in addition to the necessary prerequisite of implementation planning, the following prerequisites are the most significant.

2.1.1 Management Commitment

Management commitment is the extent to which management personnel, especially senior management, sponsor and support implementation and continual improvement of the QMS. Instituting a QMS in an organization is a significant undertaking that calls for significant investments of time, money, and effort from all involved. The need for these resources, coupled with the natural impediment of resistance to change, is significant enough an obstacle to undermine any QMS implementation effort without the full backing of senior management. Therefore, it is critically important that senior management's commitment to quality and to implementing a QMS be secured. Senior management's support during QMS implementation is also essential for securing the buy-in and cooperation of middle and lower management in the organization.

Senior management must visibly demonstrate its commitment to quality so that management's unwavering commitment to quality is known throughout the organization. Some ways in which senior management can demonstrate its commitment to quality include:

- Emphasizing the need to meet and exceed customer expectations at company events and all-hands meetings
- Establishing an organizational quality policy
- Establishing quality objectives and including them as part of the performance objectives for all employees by tying them to employee (and department) recognition and reward incentives

- Providing continuous management support to the quality management representative and quality assurance (and/or control) department*
- Requiring monitoring of customer satisfaction levels; this includes measurement of customer satisfaction, subsequent corrective action, and establishment of goals for continued customer satisfaction improvement.
- Sponsoring quality improvement initiatives
- Requiring assessment against (or registration to) a recognized quality management system standard, or application for local, state, or national quality awards (such as the Malcolm Baldrige National Quality Award)

Management's commitment to quality must not end with the successful implementation of a QMS. Sustained senior management commitment beyond implementation of the QMS is essential for an organization to continually improve itself. Keep in mind that a successful QMS implementation merely provides an organization with an infrastructure that must be utilized to realize benefits in terms of improved quality of product and services. Therefore, in order to reap the true benefits of implementing a QMS, which are long-term in nature, continued management support for the QMS is vital.

To sell quality to senior management, the quality management representative (explained later in this section) must demonstrate a need for and benefits of implementing a QMS for each of the key stakeholders of the organization — shareholders, customers, and employees. Insofar as an organization's shareholders are concerned, they are most interested in the organization's financial performance, specifically robust top-line growth and improved bottom-line performance. Need for improved bottom-line performance translates into management need to drive organizational efficiency. Therefore, a business case for implementing a QMS can be made in terms of *cost of poor quality*, which is a significant part of the total product development cost (see Table 1). For example, presenting data to senior management on how much time, money, and effort

---

* In this book, quality assurance (and/or control) department refers to the *quality function* in the organization, regardless of how it is staffed and structured. For example, this may be a separate department comprising full-time quality assurance personnel, or it may be a work group comprising personnel assigned part-time. In a small organization, it may comprise one person with full-time or part-time responsibility for this role. When feasible, the quality function should be staffed with qualified permanent (full-time) quality and/or process engineers. The quality function should be independent of and free from influence of all departments and personnel in the organization, and it should have a direct line of reporting to senior management.

Table 1 Total Product Development Cost, Cost of Quality, and Cost of Poor Quality

Total Product Development Cost = Performance Cost + Cost of Quality

Cost	*Explanation*
Performance cost	Cost of executing processes and steps that are necessary to deliver the product
Cost of quality	Cost of meeting requirements + cost of not meeting requirements (cost of poor quality)
Cost of meeting requirements	Prevention costs (for example, training, methods, tools, and procedures) + appraisal costs (for example, testing, inspections, and audits)
Cost of *not* meeting requirements (cost of poor quality)	Internal failure costs (for example, rework, re-review, and retest) + external failure costs (for example, warranty, liability, and loss of reputation)

is expended due to poor quality (which directly hurts the organization's bottom line) serves as a strong incentive for management to control cost of poor quality.

Senior management should be educated on how lack of adequate quality assurance and quality control activities directly impacts the quality of the product and services delivered, and thus negatively impacts customer satisfaction. Conversely, senior management also should be presented data from other companies that have successfully instituted a QMS and have published customer satisfaction survey data showing improved customer satisfaction levels after implementing a QMS. For example, many companies regard achievement of registration to a QMS standard, such as ISO 9000, a major milestone in their quality improvement efforts, and measure its impact (and that of other quality initiatives) on overall customer satisfaction levels.

Senior management also should be educated on how lack of processes and procedures, lack of employee training or inadequate employee training, lack of job descriptions, and unclear allocation of responsibilities contribute to employee dissatisfaction. For example, if the organization already conducts an employee satisfaction survey, employee feedback in free-text format can be collected on the aforementioned items, and analyzed to formulate corrective actions plans in these areas. Similarly, workforce maturity frameworks such as the People Capability Maturity Model (P-CMM®) can be leveraged to institute mature workforce practices, many of which also fall in the domain of (or are synergistic with) quality assurance practices.

2.1.2 Quality Management Representative and Change Agents

The quality management representative (also called the management representative), as the name suggests, is a member of the management team of an organization, and has ultimate responsibility for the definition, deployment, and continuous improvement of the QMS. Typically, the organization's senior quality officer, such as Vice President of Quality or Director of Quality, fills such a role. During QMS implementation, the management representative leads and directs the QMS implementation team, while after QMS implementation the responsibility pertains to maintenance and continuous improvement of the QMS.

A management representative should have a sound understanding of organizational processes (or knowledge from past experience with equivalent processes at another organization). He should have prior experience from having led QMS implementation successfully in other organizations (within the same industry). He should have expertise in quality, and sound project management and people management skills. The last includes an ability to lead and motivate staff members. He should have demonstrated ability to effect change in an organization by persuading and reasoning with the employees to explain the need for the change (and by explaining deficiencies in the current approach). He also should be able to represent the organization effectively in meetings with customers and other external parties.

The management representative is an organization's foremost change agent, and is assisted by other change agents, who may be full-time quality assurance personnel who are part of a dedicated quality assurance department or a cross-functional team comprising representatives who are assigned part-time to facilitate implementation of the QMS. Typically, members of such a cross-functional team participate in a permanent body that meets regularly, with the initial mandate to facilitate the definition and deployment of a QMS, and the ultimate objective to facilitate continuous improvement of the QMS (refer to the explanation of Process Management Council later in this chapter).

The change agents should possess many of the same attributes as the management representative, including adequate subject matter expertise, adequate knowledge of relevant processes of the QMS, and an ability to persuade and effect change in their respective organizations without being obnoxious or confrontational.

2.1.3 Employee Participation

Employee participation is the extent to which employees participate in the implementation, maintenance, and continuous improvement of the QMS. It is essential to secure the participation of employees, because they are the practitioners who execute the processes in the organization and thus are intimately familiar with the processes and their strengths and weaknesses. They are the subject matter experts whose acceptance of the defined QMS is vital for its deployment, use, and continuous improvement.

In *Overcoming the Improvement Paradox*, Keating *et al.* emphasize that in any improvement program, management push, though essential, has severe limitations and is unable to sustain change over the long term. The ultimate objective is to attain a self-sustaining state in which complex challenges are tackled by competent and intrinsically motivated employees. This state can be attained only when employee pull is the operative sustaining force [Keating]. In order to promote employee pull, it is necessary to involve employees in the definition of the QMS, as opposed to imposing on them a QMS that was defined in isolation, without eliciting their input. Such a QMS is perceived by employees to be alien and nonrepresentative of the actual work processes, thus increasing the odds against its acceptance.

Employee participation in QMS implementation helps secure the buy-in of the employees and reduces resistance to change barriers during the deployment phase.* By participating in the definition of the QMS, employees are able to review the QMS and offer their input during its definition. A QMS founded on the collective expertise of the employees fosters a sense of ownership and commitment to the defined system, which is essential for its acceptance.

2.2 IMPLEMENTATION PLAN

A QMS implementation is a significant undertaking with a specific goal, allocated resources, and specified time frame for implementation. A QMS implementation is no different from any other project executed in a company. It should therefore be planned, tracked, and controlled as any other project. A sample outline of a QMS implementation project plan is provided in Appendix A. Key elements of the project plan are explained in the following subsections.

* Although due to a natural human tendency to resist change, such barriers cannot be completely eliminated.

2.2.1 Implementation Goal

The first step in launching a QMS implementation project is the formulation of a clear and concise goal statement by senior management. In *Resolving the Process Paradox*, Robert Gardner cautions that the absence of a clear purpose when launching a process improvement program* can be a fatal flaw. Without purpose, there is no framework for establishing priorities, aligning efforts, or judging success. Therefore, before launching your improvement program, take the time to understand the nature and magnitude of the performance issues you are seeking to resolve [Gar01].

The goal statement should be unambiguous regarding the objectives, and it should state explicitly how the achievement of the objectives would be assessed. To this end, the defined goal should meet the SMART** criteria (Specific, Measurable, Acceptable, Realistic, Timebound):

- A *specific* goal is one that clearly states the final objective and its scope. For example, "Implement a QMS" would not be a specific goal, as it is unclear what the scope of the QMS implementation is. Does the QMS implementation encompass all products and locations of the organization? Does the implementation include design, development, production, installation, and servicing of the products?
- A goal is *measurable* if it is clear to the reader how the attainment of the goal will be validated. For example, "Implement a QMS covering the design, development, production, installation, and servicing of all products of ABC Corporation" would be a specific but nonmeasurable goal. This is because it is unclear how the achievement of this goal will be demonstrated. For example, does the organization intend to achieve registration to a QMS standard, such as ISO9000, QS9000, AS9000, TL9000, or other? (In which case this should be specified in the goal statement.)
- A goal is *acceptable* if it is agreeable to those who have been tasked with achieving the goal. If a project's goal has been foisted on a reluctant implementation team, the project is doomed to failure.
- A goal is *realistic* if it is achievable within the identified time schedule. A goal that is overly ambitious at the outset is bound to

* Note that here, QMS implementation may be regarded as a process improvement program because inherent in the implementation of the QMS is an examination of the current system and deployment of process controls and quality practices (where necessary), resulting in process improvement (refer to Chapter 5). However, sustained improvement will result only from use of continual improvement mechanisms, some of which are described in Chapter 8.

** This assumes that the goal is realistic and acceptable to the implementation team.

have little or no likelihood of being achieved. For example, if an organization with 500 employees were implementing a QMS from scratch, a goal of completing the implementation in 6 months would be nearly impossible.

- Finally, a goal is *timebound* if it clearly states by when the goal is to be achieved.

Keeping in mind the above criteria, a SMART goal would be:*

"To implement a QMS by December 16, 2005 for all activities pertaining to the design, development, production, installation and servicing of X product at all company sites of ABC Corporation. This project will be considered complete once the QMS is assessed by independent third-party quality assessors/auditors and is certified to be compliant with the defined and documented QMS."

In order to maximize the chances for the achievement of the stated project goal, it is recommended that the achievement of the goal and of the intermediate milestones be tied to annual employee objectives as part of the employee performance appraisal process for all employees. Because this affects the bonus/reward disbursement to employees, it serves as an excellent incentive to facilitate the achievement of the identified milestones and final project goal.

2.2.2 Implementation Team

One of the first tasks in planning a QMS implementation is establishment of an implementation team (see Figure 5). The QMS implementation team plans and executes the QMS implementation project and is led and directed by a project manager who reports to the QMS Steering Group. Generally, the organization's management representative serves as the project manager. (Role and responsibilities of the management representative were described earlier in this chapter.)

2.2.2.1 QMS Steering Group

The QMS steering group comprises senior management personnel from all functional areas in the scope of the QMS being defined. It exercises management oversight over the QMS. Because no functional area legitimately can claim to be exempt from defining its processes and instituting quality practices, the steering group should comprise senior management from all functional areas in the organization. The senior management

* This assumes that the goal is realistic and acceptable to the implementation team.

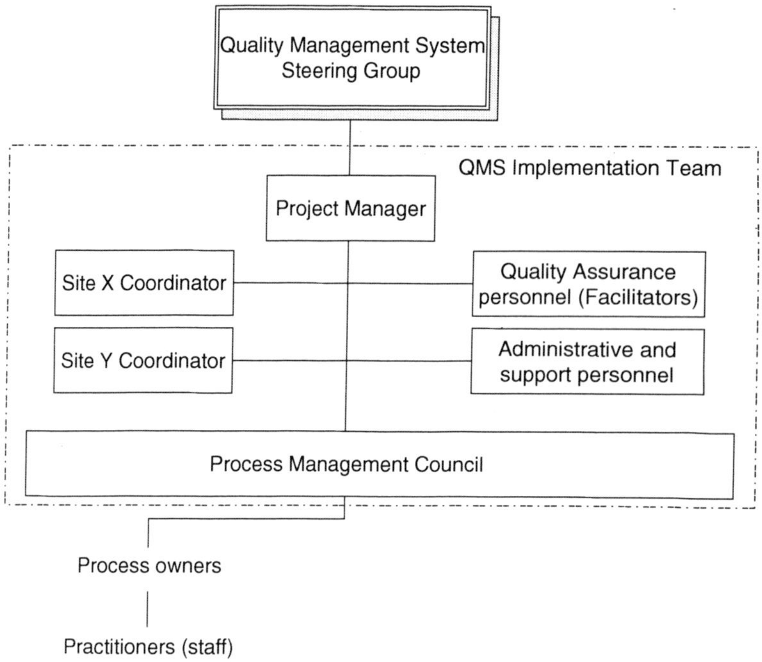

Figure 5 Organization Chart of QMS Implementation Team

person who has overall responsibility for the organization (or business entity) for which the QMS is being established, and who is the project sponsor for the effort, generally heads the group. This body should be a permanent entity, which should remain intact even after QMS implementation. The steering group has the responsibility to:

- Set strategic direction and exercise strategic decision making with regards to quality
- Establish long- and short-term quality objectives
- Act as the sponsor and provide needed resources to implement, maintain, and continually improve the QMS
- Direct and oversee implementation and continual improvement of the QMS
- Emphasize, demonstrate, and communicate to employees management's commitment to quality
- Provide timely management intervention, when necessary
- Conduct periodic management review of the QMS (refer to further discussion in Chapter 5)

2.2.2.2 *Process Management Council*

As explained earlier in this chapter, it is necessary to involve employees in the QMS implementation effort. The process management council (PMC) is an effective vehicle for this purpose. It is a cross-functional team comprising representatives from all functional areas in the scope of the QMS implementation. Management personnel from each relevant area should consult with the project manager to appoint a representative to participate in the PMC. By providing resources for and continually supporting the activities of the PMC, senior management of each functional area can demonstrate visibly the pledged management commitment to the QMS implementation effort.

There are certain desired qualifications of a PMC member that are described next. A PMC member should:

- Be recognized in the organization as a person who is committed to quality. The project manager should be confident that this person would be a valuable contribution to the implementation team.
- Ideally be an employee who has been with the organization for a sufficient amount of time that he or she is knowledgeable in the department's processes and in other cross-functional processes that involve the department
- Have extensive and varied industry experience in the respective functional area
- Have good oral and written communication skills
- Not be a person who is regarded as the most critical employee of the department and who already has a significant workload. This is because a QMS implementation requires a significant amount of effort from the implementation team; lack of satisfactory participation by all project participants can affect the QMS implementation progress in the respective area significantly.

PMC members participate on an as-needed basis in all activities and meetings pertaining to the QMS implementation; they generally meet periodically, and such meetings are chaired by the project manager (refer to Section 2.4.1). In the case of large organizations, a PMC member also may represent all the subdepartments within that department, and separate PMC members for each of the subdepartments may not be required. Responsibilities for PMC members may include but are not limited to:

- Serving as the single point of contact between the quality assurance department and the PMC's department for all issues pertaining to

the QMS implementation (including participation in implementation planning, and championing QMS implementation in the department)
- Participating in the identification of business processes in which the PMC's department is involved
- Producing QMS documentation assigned to the PMC member by the implementation team. The PMC member also may delegate generation of some QMS documentation to other individuals in his department, but he is ultimately responsible for the assigned task being completed as per the agreed dates.
- Ensuring that interfaces with other processes and departments have been addressed in the QMS documentation and that the documentation has been reviewed with the interfacing departments. This is necessary to ensure that the documentation is correct and complete (preliminary "process verification").
- Ensuring that all requirements applicable to the process have been satisfactorily addressed (preliminary "process validation")
- Participating in the review of cross-functional QMS documentation — for example, procedures spanning multiple departments
- Providing process training (or assisting the quality assurance department to provide process training) to employees on processes owned by the PMC member's department
- Implementing corrective and preventive actions, as required, after quality audits and assessments

2.2.2.3 Site Coordinators

In organizations where QMS implementation is across multiple sites, it generally is desirable to appoint a site coordinator. The site coordinator serves as the single point of contact for:

- Communications with the management representative (or PMC)
- Tracking and reporting QMS implementation progress for the site
- Disseminating messages to the site personnel
- Arranging celebratory events at the site

Generally, one of the PMC members from the site or a member of the quality assurance department at the site may serve as a site coordinator.

2.2.2.4 Other Members of QMS Implementation Team

The QMS implementation team may comprise administrative or support personnel to assist with ancillary tasks during QMS implementation. For example, support staff generally is required for definition and implementation of QMS documentation and metrics repositories, development of in-house tools to support the QMS, and other such tasks. The need for such support tasks generally is higher during establishment of the QMS, and need for such resources declines significantly once the QMS enters a maintenance and continuous improvement phase after implementation. Therefore, it is preferable that need for such support personnel be met by securing qualified resources on a part-time basis from the appropriate departments.

Finally, the QMS implementation team also consists of quality assurance personnel who are members of the organization's quality function (as explained earlier in this chapter). In essence, the project manager and the quality assurance personnel serve as the core team within the QMS implementation team. They formulate much of the implementation approach and exercise most of the tactical decision making (with active consultation and involvement of appropriate QMS implementation team members).

Note that the responsibility for implementing the QMS does not belong to the QMS implementation team alone. Quality is after all the responsibility of every employee in the organization. The QMS implementation team is assisted by process owners and process practitioners who are called upon to contribute their expertise on an as-needed basis. These are described next.

2.2.2.5 Process Owners

Every process has a designated process owner, generally a management person. Process ownership typically is assigned to the department in which most of the process activities are performed, or the department that has the maximum stake in the correct execution of a process. Some process owners may serve as PMC members, while others may prefer to work with their respective PMC representative for defining their processes. In order to promote cross-functional awareness and discussion on processes between appropriate people, the management representative should ensure that the list of processes and associated process owners is known and communicated throughout the organization. This is especially important during QMS implementation, when QMS documentation is still being developed and pertinent information on each business process and the respective point of contact is not readily available.

The responsibilities of process owner include but are not limited to:

Defining and documenting the process for which the person is responsible. In case of cross-functional processes, this includes responsibility for requesting and securing needed resources (personnel) for this purpose from other departments that are involved in the process.

In this case, the process owner is also responsible for cross-functional review of the defined process.

Acting as the communication interface between the process and functional (or line) management. In this context, the latter has the responsibility for process execution and adherence, and for requesting process changes or improvements from the process owner, as necessary.

Clarifying interfaces with other processes

Establishing controls to monitor process execution

Establishing measurement points in the process to measure process effectiveness against goals

Providing employee education and training on the defined process

Monitoring that the process continues to meet the needs of its customers (internal and external)

Taking timely corrective and preventive action, when required

Continually seeking opportunities to improve the process; this includes handling of improvement suggestions in consultation with the quality assurance department.

2.2.2.6 *Process Practitioners*

Process practitioners are personnel who execute a particular process. Therefore, in essence, every employee of an organization is a process practitioner as well. During QMS implementation, they contribute their subject matter expertise (as needed) to assist the process owner and/or PMC member in performing their duties. For example, appropriate process practitioners should be invited to participate in the definition, documentation, and review of a process, when necessary. A process owner or PMC member also may delegate the task of writing a procedure to a suitable process practitioner.

2.2.3 Implementation Strategy

2.2.3.1 *What does a QMS Implementation Entail?*

Recently, there has been a gradual but noticeable paradigm shift in the quality field. Increasingly, organizations are beginning to realize that the only way to understand how an organization operates, and thus to improve its performance, is by discussing and assessing quality in the context of business processes. This has resulted from a growing realization of the obvious — that all work in an organization is accomplished by the execution of business processes. Therefore, the quality of a product is largely dictated by the capability of the process used to produce it. That is, a high-quality process will deliver a high-quality output. This realization has led to a shift in management style from department or function management to process management. Consequently, process management and process improvement often are used synonymously with quality management and quality improvement, respectively.

Due to this paradigm shift in the quality profession, the task of establishing a QMS is viewed fundamentally as the establishment of the necessary infrastructure to manage and continuously improve an organization's processes. Such a process management infrastructure comprises the organization's process assets. This does not necessarily mean that the primary process asset — organizational processes and subprocesses — are documented. The decision to document a process and the extent to which it needs to be documented (by creating procedures and work instructions) are based on organizational need, and are dependent upon factors such as those described in Section 3.2.2 in Chapter 3. Additional process assets include tools, methods, resources, employee competencies, guidelines (for execution and/or tailoring of defined processes), checklists, templates, forms, standards, and other such documents that are necessary to support the execution of the organization's defined processes. Once personnel start executing the processes in the defined QMS, the documents and records produced serve as evidence of use the QMS.

2.2.3.2 How to Proceed with Implementation: Top-Down or Bottom-Up?

Because organizational processes are the most significant part of an organization's process management infrastructure, they should be the first to be defined. In order to define an organization's processes, they must first be identified. The concept of process mapping, which is explained in Chapter 4, helps stimulate initial discussion on *how* an organization transforms inputs into outputs by the interplay of various organizational processes.

In essence, this sequence of business processes (and activities) is a value chain wherein each process (or activity) has a specific purpose and added value. Process mapping helps construct a process-oriented model of an organization's business operations (or value chain). As the processes and their interdependencies are iteratively mapped and refined, there is an improved understanding of the purpose and scope of each process, its content, and the added value it delivers within the overall context. This exercise results in the formulation of an initial coarse outline of an organization's high-level processes that subsequently can be elaborated upon (to the extent necessary).

As is intuitively obvious from the aforementioned description, the task of process mapping must logically proceed in a top-down (as opposed to bottom-up) manner. That is, initially, upper management should map business processes at the highest level of abstraction in the organization. The management representative and/or quality assurance personnel should facilitate this exercise. This should be followed by decomposition and elaboration of each higher-level process into underlying subprocesses by mid-level and lower-level management personnel and practitioners. Note that this need for increased process definition (and documentation) is necessitated by the increase in the technicality and complexity of processes at lower levels of the organization, and the direct bearing they have on product and service quality.

As the processes are defined in greater detail at lower levels of the organization, the need for supporting process assets increases correspondingly (see Figure 6). Availability of adequate process assets provides the necessary safeguards to minimize the likelihood of process breakdown by providing the greatest infrastructural support to the practitioners at the level where much of the organization's work is accomplished. Again, it is intuitively obvious that to define process assets, a need for them must first be identified. That is, one must take a high-level view of each process, its scope, and its content, and carefully examine what process assets are necessary to support and demonstrate the consistent faithful execution of the process.

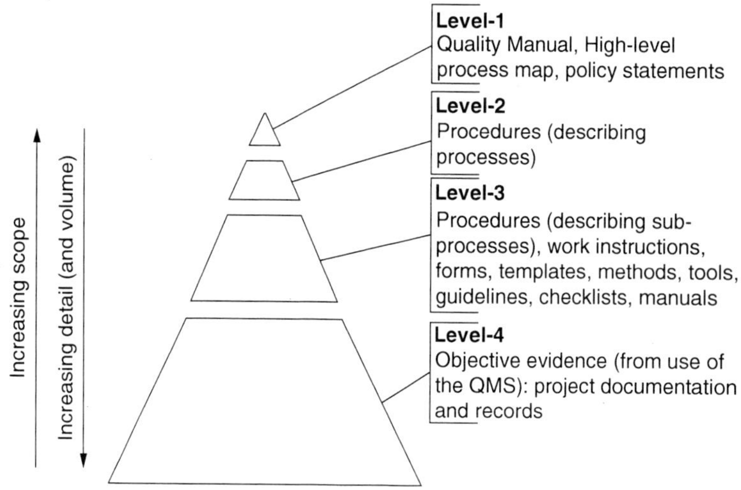

Figure 6 QMS Documentation Hierarchy

A top-down approach to QMS implementation helps an organization develop a systems view of the organization by first looking at the business processes as a whole and then gradually peeling the onion and looking inside each process. The organization progresses from recognizing and examining interdepartmental processes and subprocesses (processes and subprocesses between departments) to intradepartmental processes and subprocesses (processes and subprocesses within departments).

On the other hand, a bottom-up approach to QMS implementation entails identifying all the day-to-day tasks performed in the organization (within the scope of the QMS) and creating QMS documents to describe them. Such an implementation approach results in excessive and haphazard creation of processes and process assets because of disjointed efforts of various personnel. This lack of coherence and well-reasoned creation of processes and process assets is due to inadequate understanding of the relevance of each process (or activity) in the overall context. This is because a bottom-up QMS implementation prevents an organization from seeing the big picture — the overall value chain of the organization. That is, it inhibits the organization from developing a system-level view of its business processes — a view that is essential to understanding and improving complex business operations. Instead, it promotes a functional view of the organization as each department attempts to define its processes and process assets, and begins optimizing such intradepartmental processes, often at the expense of interdepartmental and overall organizational performance* (see Figure 7). Because most business processes are cross-functional, implementation of significant improvements in an organization often involves analyzing and improving the interfaces *between* departments and business processes.

In *The Process Edge*, Peter Keen says it is not uncommon to see organizations claim many successful improvement initiatives while overall performance fails to improve [Kee97]. Keen contends this phenomenon occurs when organizations focus their improvement efforts on relatively unimportant processes. In essence, the organizations either fail to recognize their core value-creating processes (due to lack of awareness of the big picture) or they indulge in functional (or local) optimization — both of which result in little if any overall improvement.

To summarize, local optimization at the expense of system-wide optimization defeats the very purpose of instituting a QMS: to improve organizational performance across all functions and at all hierarchical levels in the organization. For the aforementioned reasons, QMS implementation must proceed in a top-down fashion, and the need for each

* Refer to previous discussion in the Preface.

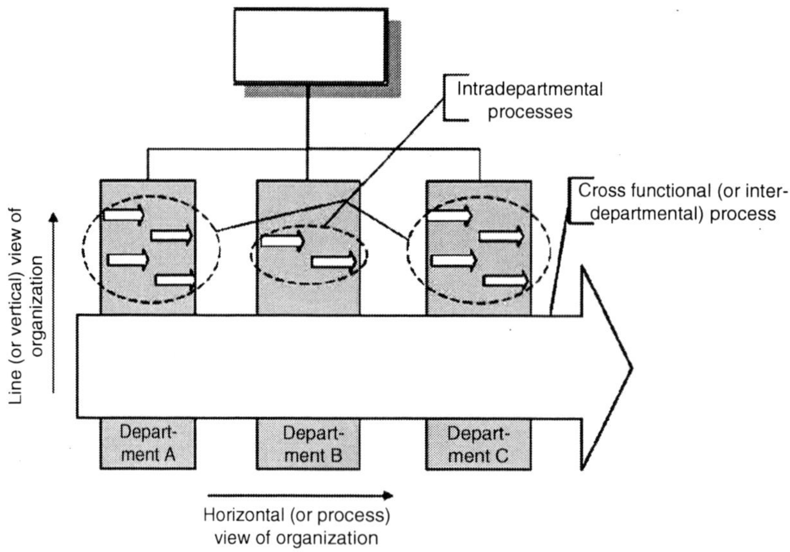

Figure 7 Functional vs. Process View of an Organization

process asset and the benefit it provides must be examined carefully in the overall context.

2.2.4 Implementation Process

The overall QMS implementation approach comprises two elements:

1. High-level implementation strategy (explained in the previous section); and
2. The actual process to be followed to implement the system.

The process that most organizations follow for implementing a QMS is similar to the Deming cycle explained in Chapter 1.

By following a structured process for QMS implementation, where the implementation progresses through distinct phases, each with its unique purpose, an organization is able to take a piecemeal approach to QMS implementation. The achievement of the end-goal is planned by decomposing it into a series of feasible intermediate goals that must be met in each implementation phase. The benefit of this piecemeal implementation approach is that it enables the QMS implementation team to remain focused throughout the course of the implementation by driving towards the achievement of near-term goals that are clearly defined and contributory to the achievement of the end-goal. Furthermore, such an approach provides structure and purpose to each stage of QMS implementation. This enables the implementation team to establish short-term priorities, align effort, assess daily progress, and take corrective action in a timely manner. It also enables the implementation team to maintain and uplift employee morale and employee participation by demonstrating progress towards the end-goal, and it provides opportunities to celebrate the achievement of the intermediate milestones.

The QMS implementation phases are as follows (see Figure 8):

First, the QMS implementation is planned (*Plan* phase).
Next, the QMS is defined and documented (*Do* phase).
The defined QMS then is verified and validated to ensure it meets applicable requirements (*Check* phase).
The QMS then is deployed across the organization, and continuously improved (*Act* phase).

This is not to imply that the phases are strictly sequential. On the contrary, they are iterative, and there is feedback from one phase to the previous phases. For example, as some QMS processes are defined and documented, they may be reviewed, approved, and released for use by personnel. Subsequently they may be rereviewed and refined (once interfacing processes have been defined), and reapproved and rereleased for use. In other words, the QMS definition and deployment is evolutionary. As the QMS gradually is rolled out, personnel become accustomed to the new way of working within a formal QMS, as opposed to being subjected to sudden and drastic change when the QMS is "turned on" abruptly.

Each of the QMS implementation phases may be formally defined as follows:

QMS planning phase (Phase I of this book) — The QMS planning phase entails the specification of the QMS implementation goal, and lays out the roadmap that the organization will follow to achieve the defined goal. Plans devised at this stage may be revised later (as needed), but for the most part, decisions made in this phase will have a profound effect on the pace and thoroughness of the QMS implementation effort, and quality of the resulting QMS. It is therefore important that there be adequate forethought and meticulous planning* in this phase (covering related items discussed in this chapter and in Chapter 3). These items are summarized below:

* Most of the output of the planning exercise will be captured in the project plan for QMS implementation, a sample of which is provided in Appendix A.

Implementation prerequisites: Identify prerequisites for success, and ensure that they have been secured. Doing so helps maximize the chances of success and helps mitigate risks from the beginning of the QMS implementation.

Implementation goal: Establish a clear goal statement that satisfies the SMART criteria explained earlier in this chapter.

Implementation team: Plan for an implementation team with adequate cross-functional representation and clearly defined roles and responsibilities that are communicated to all implementation team members.

Implementation strategy: Brainstorm the implementation strategy and ensure that it is clearly communicated to all implementation team members and staff members.

Implementation process: Define an implementation process that lays out the high-level roadmap for implementing the QMS. This includes main phases of QMS implementation and key activities within each phase (refer to the sample roadmap shown in Figure 8).

Implementation schedule, needed resources, and cost: For the implementation roadmap defined in the previous step, estimate the resources for each activity and establish an implementation schedule. Ensure that an adequate contingency is included within each phase. Also, estimate the major expenses for the implementation and budget for the anticipated implementation costs.

Mechanisms to manage the implementation, communicate progress, and encourage employee participation: Identify the mechanisms that will be used to track and control implementation progress to ensure that the project progresses according to plans. Identify means that will be used to communicate progress, especially to senior management, and to facilitate timely management intervention in case progress lags. Identify means that will be used to encourage employee participation and recognize employee contribution.

QMS documentation: Ensure that the key elements of a sound QMS documentation management system are in place. Ensure that the process for creating, reviewing, and approving new QMS documentation is defined and communicated to all employees. Both these items are discussed in detail in Chapter 3.

QMS definition phase (Phase II of this book) — The QMS definition phase entails the definition and documentation of the organization's QMS. If the organization has selected a quality management system standard for use, this phase entails the definition of the QMS in accordance with the standard's requirements. Activities in this phase include but are not limited to:

1. *Requirements analysis (if applicable)*: Analyze each requirement in the QMS standard to clearly understand what is required and how the requirement can be satisfied.
2. *Gap analysis*: Assess the current state of the system (processes and procedures) against best practices in the given industry, or against requirements in the QMS standard (if one was selected). This exercise is intended to obtain answers to the question *Where are we right now?* With a better understanding of where one is and where one is headed (project goal), one is better able to plan future action.

   The gap analysis may reveal that processes and procedures are already in place in certain areas of the organization. When appropriate, reuse all or part of the current implementation as opposed to beginning from scratch. The gap analysis may also reveal critical quality discrepancies requiring immediate attention; these should be planned for immediate rectification (see step 4).
3. *Revise implementation plan*: As is obvious from the foregoing explanation, the gap analysis typically will cause the implementation plan to be fine-tuned as per the insight gained into the current state of the system.
4. *Correct critical quality (or process) deficiencies*: Act upon the results of the gap analysis to correct critical quality (or process) deficiencies that can be fixed relatively easily. Doing so provides immediate return on investment for initiating the quality implementation effort. It also provides an opportunity to cite the success stories to sustain management commitment and encourage employee participation.
5. *Perform high-level process mapping and create supporting process documentation*: Perform process mapping for high-level organizational processes, and create supporting process documentation, as needed.
6. *Perform low-level process flowcharting and create supporting process documentation*: Perform process flowcharting for lower-level organizational processes, and create supporting process documentation as needed.
7. *Create additional QMS documentation:* In addition to QMS documentation in the form of process maps, create additional documentation as needed, such as procedures, templates, and forms.

Notes (applicable to steps 5, 6, and 7):

Note 1: Refer to related guidance on quality practices relevant to different functional areas of an organization in Chapter 5. (This will have a bearing on the content of the process maps, and supporting QMS documentation and process assets.)

Note 2: When the organization is implementing a QMS in accordance with a QMS standard, address those requirements that are not being addressed currently, and enhance current implementation in cases where the current implementation is partially compliant with the standard's requirements.

Note 3: It is recommended that required QMS documentation be created in a prioritized manner. For example, the internal audits procedure may be created after the relatively higher-priority task of documenting product development processes has been completed. As each requirement is implemented and associated QMS documentation is prepared, it should be verified for correctness, completeness, and consistency, and validated for compliance to internal quality requirements (and QMS standard requirements, if applicable). Such verification and validation should be performed by conducting peer reviews with appropriate personnel to ensure that each QMS document is correctly defined and satisfies applicable quality requirements.

QMS refinement phase (Phase III of this book) — The QMS refinement phase involves a final verification of the entire QMS to ensure that all processes interact as originally planned and, further, that all processes are mutually consistent and correctly defined. This phase also involves a final validation to ensure all elements of the QMS comply with the organization's quality requirements (and, if applicable, requirements of the QMS standard in use). Deficiencies identified during this phase are addressed by requesting corrective action from the respective process (or document) owners.

QMS deployment phase (Phase IV of this book) — The QMS deployment phase involves institutionalizing the QMS across the organization. In this phase, the QMS is rolled out incrementally so that it gradually is adopted and becomes the new way of working. As each process is defined, documented, and approved for use by employees, it enters the QMS deployment phase. In this phase, employees are trained on the defined processes, and execution of the processes is monitored by the quality assurance personnel (and by QMS implementation team members, as appropriate) who participate in or observe activities as they are executed. They verify that processes are being executed correctly and that they are adequate and effective. Process execution also is verified by means of internal quality audits performed during or after process execution.

This does not imply that during QMS deployment, all old processes are thrown out and replaced with new high-quality processes. However, establishment of a QMS will cause the organization to examine all its existing processes, and it is safe to assume that most of them would be impacted to some extent during QMS implementation (mostly in terms of needed improvements). This examination also may reveal some inadequate and inefficient processes that need to be discarded and replaced with new ones. For the most part, however, implementation of the QMS generally will result in changes to existing processes, with some processes undergoing major change and others undergoing minor change.

It is important to plan for a certain amount of time (typically, months) between completion of employee training and commencement of internal quality audits. This time period, referred to as the process establishment period, is required for two reasons: first, some amount of time is required to adequately promulgate a QMS throughout an organization such that it becomes well entrenched in the organization by becoming an inherent part of how the organization conducts everyday business. Second, some amount of time is required to build sufficient amount of evidence of use of the QMS that then can be audited. Starting an internal quality audit program too soon might not provide internal auditors sufficient evidence that is necessary to assess adequacy and effectiveness of the QMS.

The process establishment period varies from one organization to another and depends on factors such as:

- Lead time to develop the products (in order to allow all product development processes to be exercised at least once);
- Extent of change to current processes; and
- Amount of QMS training provided to employees.

It is to be expected that, beginning with the process establishment period, the organization will encounter growing (maturing) pains because the natural human tendency to resist change will begin to surface once the QMS begins to directly affect how everyone does their work. Further, employees and management personnel unaccustomed to the new and/or refined processes in certain cases will attempt to circumvent the process when operational and schedule considerations overtake quality considerations.

In *10 Process Improvement Lessons for Leaders*, Gardner states that in the short-term, the aforementioned dynamic places the organization under increased stress because the cost* of an improvement effort is immediate, while most of the benefits generally are delayed. That is, in the short-term, employees and management might be able to see few benefits, while

* Cost here includes need for additional time and effort due to the learning curve associated with new or refined processes, as well as costs incurred from additional quality practices now required to be executed.

many of the benefits typically would be realized (and validated by measurement data) over the long term. (QMS implementation benefits were explained in Chapter 1.) He warns that QMS implementation teams should resist pressures to show quick results and guard against a tendency to manipulate and exaggerate results in the short term to look good and demonstrate immediate return on investment (because the latter generally is not realized in the short term). Such an approach promotes form over substance, and once the credibility of the QMS is suspect, its effectiveness and survivability are seriously threatened.

The QMS implementation team will need to address the aforementioned challenges by continually educating management and personnel (and managing their expectations), emphasizing benefits already realized, monitoring process execution (as described earlier), and reasoning with personnel and working cooperatively with them to overcome resistance to change.

Continual improvement phase (Phase V of this book) — With the completion of the QMS deployment phase, an organization effectively transitions to a state where compliance with the defined QMS needs to be continually monitored and the defined system needs to be continuously improved and optimized. This is the final and never-ending phase of QMS implementation — the continual improvement phase. It entails the use of mechanisms necessary to facilitate continuous improvement of the QMS.

Mechanisms for continuous improvement are not necessarily established in this phase only. Some of them may have been defined in the QMS definition phase and deployed in the QMS deployment phase. Others may have been defined but their deployment deferred until the continuous improvement phase. Yet other continuous improvement mechanisms may remain undefined until this phase. For example, an organization typically does not start collecting customer satisfaction data until a few months after completion of the QMS deployment phase, else it would be too early for its customers to realize benefits from the implementation of the QMS. Moreover, collecting customer satisfaction data too soon might not provide readily actionable information because many of the known deficiencies might be attributed to causes that the organization already is addressing under the QMS implementation underway.

Mechanisms for continuous improvement include but are not limited to:

Metrics program for process and product improvements — This entails the use of process, project, and product metrics* to improve the quality of existing processes and products continually. It enables the organization to quantitatively drive process and product improvements by measuring current performance and establishing quantitative objectives (and supporting plans) for improvements.

Quality objectives — These entail the establishment of long- and short-term quality objectives (as explained in Chapter 1) and formulation of supporting improvement plans for the achievement of the identified quality objectives.

Internal quality audits — Internal quality audits must be utilized not merely to verify adherence to the defined QMS but also to explore opportunities for continuous improvement. In order to serve as a vehicle for continuous improvement, internal quality audits must go beyond mere compliance auditing. Internal auditors must identify situations that require preventive action, and draw upon their experience to make observations and suggestions regarding possible improvements. Internal auditors also should help disseminate best practices from within the organization so that other organizational functions can adopt the best practice, if appropriate. This is especially relevant in large organizations or organizations with multiple locations.

Corrective and preventive action requests — This entails the establishment of mechanisms that the quality assurance personnel can use to request corrective and preventive actions from various departments, as needed.

Lessons learned from past projects — Postmortem analysis of past projects can provide valuable information on positive experiences that need to be replicated in future projects, as well as opportunities for improvement. In this way, as the organization permanently deploys positive experiences and begins addressing the areas identified for improvement, it is able to continually improve its QMS. (Refer to Appendix C.6.)

Customer satisfaction data — It is widely accepted that if an organization's QMS is adequate and effective, it generally will result in satisfied customers (although customer satisfaction also is affected by several other parameters, such as product pricing). Collection and analysis of customer satisfaction data provides valuable information on the extent of customer satisfaction (or dissatisfaction), as well as areas that need improvement. Continuous monitoring of

* Such metrics generally are complemented by organizational and individual performance metrics and objectives (e.g., sales-related metrics, employee retention/attrition metrics, and so on) for improvements in organizational business performance and employee job performance. However, such metrics are outside the context of quality-related metrics, and thus are beyond the scope of this book. A recommended reference for such metrics is [Kap92].

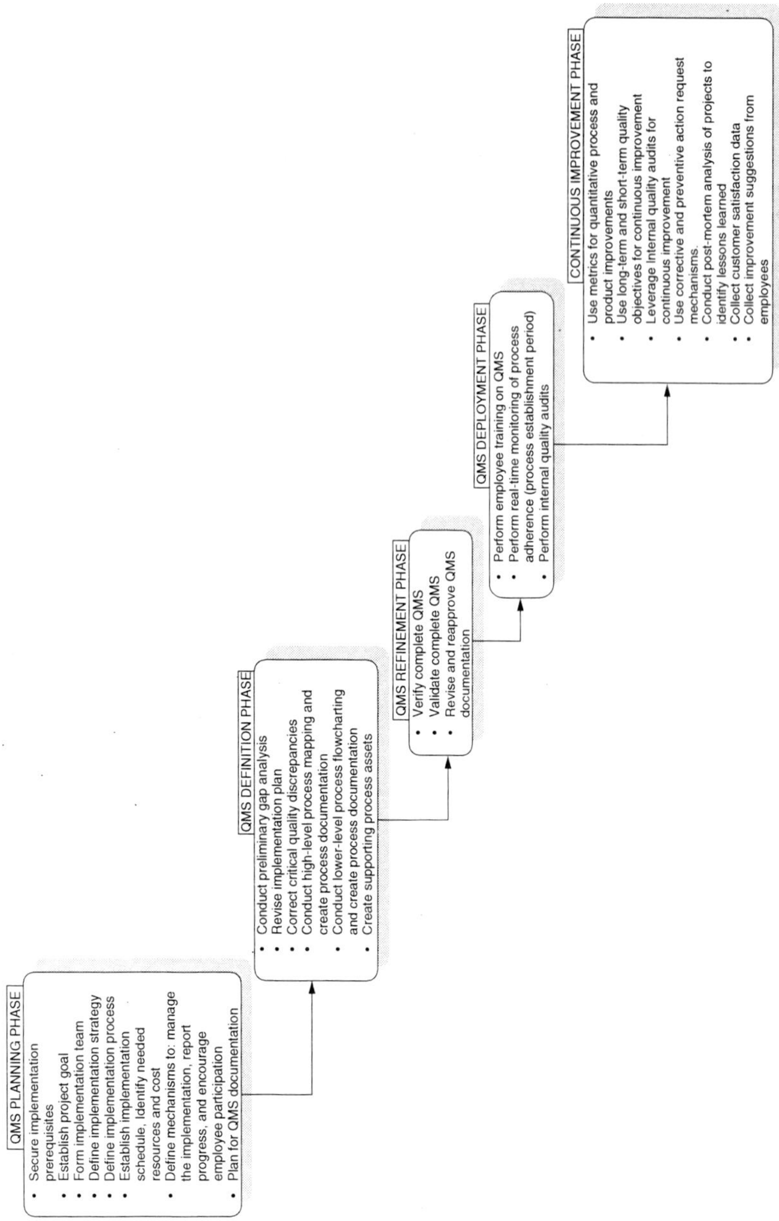

Figure 8 QMS Implementation Road Map

customer satisfaction data therefore serves as an effective means for driving continuous improvement in an organization.

Improvement suggestions from employees — There are perhaps no better means of improving processes than eliciting improvement suggestions directly from the practitioners of the respective processes. Process practitioners are intimately familiar with the strengths and weaknesses of their processes and therefore can provide valuable information on what works and what does not. Process practitioners are more likely to provide improvement suggestions if there is a well-known mechanism to actively elicit their suggestions, as well as employee awareness about information (or statistics) regarding implementation of past improvement suggestions.

2.2.5 Implementation Cost

The cost of implementing a QMS varies from one organization to another because the costs associated with the various factors that determine overall implementation cost are unique to every organization. As a result, it is not possible to provide a cost estimate of what an organization realistically can expect to spend for implementing a QMS. However, it is important for organizations to be aware of the factors that influence cost of QMS implementation, and plan accordingly. Admittedly, the overall time to implement the QMS (discussed in the next section) also will influence the overall cost of implementation; implementation cost and implementation time frame are, after all, closely related. Following is a list of factors that directly affect implementation cost:

Personnel costs — There is a significant amount of cost associated with hiring and retaining personnel who have the relevant industry experience in QMS implementation (and its subsequent maintenance and continual improvement). If an organization decides to engage the services of a consultant for QMS implementation, there are consultancy costs that need to be budgeted. Personnel costs also include the cost of employee participation in the project (although such costs are typically not budgeted). It is beneficial to estimate the time required for QMS implementation tasks from various employees so that management personnel can plan for its expected impact on the regular job responsibilities of the employees, and the QMS implementation team can take it into consideration while establishing the implementation schedule (refer to related discussion in the next section).

Tool costs — The tools (such as software) an organization needs to procure to implement its QMS will have a bearing on the implementation cost. For example, software tools typically are required for preparing process documentation, electronic implementation of a QMS documentation repository, and so on.

Training costs — If an organization decides to engage the services of a consultant for training employees on quality principles and the QMS, the organization must budget for such costs. Such training costs will depend on:

- The number of employees to be trained;
- The number of training courses required; and
- Whether the training is to be conducted periodically (to account for new employees, employee transfers, and employee promotions) or is a one-time event (with future training to be provided by the organization's internal quality assurance department after appropriate knowledge transfer from external consultants).

Training costs include costs pertaining to purchase of books and other reference material for self-study purposes.

Travel costs (if applicable) — In the case of multi-site organizations, travel-related costs for the purpose of intersite coordination, consistency, and communication comprise a significant amount of the total implementation cost.

Professional memberships and networking events — Organizations should budget for costs associated with professional memberships, which can be a valuable resource for acquiring subject matter expertise. Organizations also should budget for costs associated with attending networking and industry events (or sponsoring such events). Such interaction with peers from other organizations provides valuable implementation insights from other organizations, and helps in sharing of lessons learned.

Employee recognition rewards — Organizations must budget for costs associated with recognizing employee contribution by distributing employee recognition awards, celebrating achievement of project milestones, quality giveaways, and ongoing contests to promote awareness of the QMS.

Registration cost (if applicable) — In the case of organizations seeking registration to a QMS standard, the organization must budget for registration costs. Such costs include but are not limited to cost of preassessment audit (and/or gap analysis audit) by the registrar, documentation audit, registration audit, surveillance audits, and travel costs (of the registrar to the company's locations). All of these costs are a function of the number of days that will be required to audit the organization, and will vary from one organization to another.*

---

* The travel costs also are a function of the number of external auditors that would be required to conduct the audit.

Registrars are authorized by a formal accreditation body to perform registrations to a particular QMS standard. Accreditation may be described briefly as an independent assessment (by a recognized body) of the registrar's competency to perform particular types of audits. Such accreditation bodies also establish guidelines regarding the number of audit days required for an organization with an employee count within a particular range. Generally, the registrar can provide a quote for the registration cost that includes the aforementioned expenses.

2.2.6 Implementation Time Frame

2.2.6.1 Factors that Influence Implementation Time Frame

Just as the cost of QMS implementation depends on various factors and varies with each organization, the time frame for implementation too is dictated by various factors and is unique to each organization. It is necessary for the QMS implementation team to be cognizant of these factors so that it can establish a realistic implementation schedule. Some of the factors that influence overall implementation time frame and, in turn, affect implementation cost are:

Implementation prerequisites — Implementation time frame depends to a significant extent on how well an organization satisfies the implementation prerequisites. If implementation prerequisites are not met, the implementation time frame will be significantly higher than if they had been met. This generally results from need for more time to implement the QMS (due to lack of sufficient management commitment and employee participation), increased rework (due to lack of adequate planning and foresight), and other such reasons. In fact, organizational inability to meet implementation prerequisites also may ultimately result in failure and gradual demise of the QMS implementation project.

Scope of the QMS and size of the organization — The scope of the QMS has a bearing on implementation time frame because it determines the number of functional areas (departments) and personnel that would be part of the QMS implementation. The greater the scope of the implementation, the longer the implementation time frame.

Current state of the system — The current state of the system, as determined from the preliminary gap analysis, provides valuable information on the anticipated implementation time frame. The fewer required QMS elements that already are implemented (and deemed satisfactory), the longer the implementation time frame, as more effort is required for QMS implementation.

Process establishment period — The time required for process establishment varies by organization, and will depend on factors described earlier in this chapter. The greater the process establishment period, the longer the overall QMS implementation time frame.

Number of required internal quality audits — The number of internal audits required to assess the QMS and gain confidence regarding its use has a direct bearing on overall QMS implementation time frame. The greater the number of internal audits and resulting corrective actions required to verify complete deployment and use of the defined QMS, the longer the implementation time frame.

2.2.6.2 Time Requirements for Implementation Tasks

In addition to awareness of the above factors that influence implementation time frame, the implementation team also should be aware of the time required for the different types of QMS implementation tasks. As stated earlier, this enables the QMS implementation team to arrive at a more realistic implementation schedule after careful consideration and aggregation of the time required to execute the underlying tasks.

Table 2 lists some of the fundamental tasks that should be considered by the QMS implementation team while planning the detailed implementation schedule (shown in Table 4). This table is not supposed to list all the tasks that are part of the implementation,* but it is meant to identify the *types* of tasks (and time required) to execute each of them. Note that the time allocation shown in Table 2 is just an example, and planned times will vary based on each organization's needs. For example, while one organization may plan for 6 person-days per year for QMS overview training, another organization may plan for 12 days per year. Similarly, an organization that already has a reasonably mature QMS in place might plan less time for mentoring of QMS implementation team members than an organization implementing a QMS for the first time.

For estimating the time required to accomplish each task, the basis of estimates should be documented so it can be explained how each estimate was arrived at. For example, if documentation of each QMS procedure is expected to require 40 person-hours' effort, then the basis of estimate may be recorded as shown in Table 3.

In cases of large implementation activities that comprise several underlying tasks, and thus several variables that may affect the total anticipated time for completion of the activity, it is recommended that risks inherent in the estimation process be minimized by asking more than one estimator to provide a task estimate. For example, the management representative, in addition to estimating the task, may request an estimate from two or three other people, such as quality assurance department personnel or an appropriate PMC member (or process owner).

* These are contained in the implementation schedule and QMS implementation tracking log described later in the chapter.

Table 2 Sample Tasks That Should Be Considered When Preparing a QMS Implementation Schedule

Implementation Task	*Time Allocation (in person-days)*
Analysis, documentation, review, rework, and approval of each QMS document, such as procedures, standards, and templates	7 (for each)
QMS Overview training for all employees (including training preparation and execution)	6 (per year)
Training for each department on applicable QMS documentation	4 (per department, per year)
Internal quality audits (including audit planning, audit preparation, audit interviews, audit reporting, and corrective/preventive action follow-up)	4 (per department)
Mentoring/consulting of personnel by quality assurance department	1 (per week)
Management review of QMS (including meeting preparation, presentation, and follow-up)	1 per month (minimally, bimonthly management review)

Table 3 Example of a Basis of Estimates

Task/Activity Name:	*Preparation of a QMS Procedure*
Assumptions	Use the 80% rule (i.e., assume an employee performs only 32 person-hours' productive work in a 40-hour workweek). Include a minimum of a 10% contingency for time required to accomplish the task.
Effort required	40 person-hours/32 person-hours = 1.25 person-week (i.e., 50 person-hours)
Contingency (10%)	5 person-hours
Final estimate	50 + 5 = 55 person-hours or approximately 7 person-days (assuming 8-hour work day)

2.2.6.3 Preparation of Implementation Schedule

Once the QMS implementation team is aware of the factors that influence implementation time frame and the approximate time required to accomplish fundamental implementation tasks, preparation of the detailed implementation schedule may begin. In order to prepare the implementation schedule, the team first will need to define a high-level roadmap for the achievement of the project goals. An example of such a roadmap was provided in Figure 8.

Each phase in the implementation schedule should comprise key activities belonging to that phase, along with due dates and people responsible, and an identified due date for phase completion (refer to Table 4). Such phase completion dates serve as project milestones to assess project progress, facilitate timely corrective and preventive action, and celebrate achievement of the intermediate objectives associated with each phase. Additional milestones can be defined to correspond to the completion of critical activities in the implementation schedule. It is recommended that the traditional Gantt chart representation of the implementation schedule also be prepared (refer to Table 5).* The implementation schedule serves as a high-level strategic work plan for QMS implementation. It should be supported by a lower-level tactical work plan for planning and tracking underlying tasks. (Refer to the explanation of the QMS implementation tracking log in the next section.)

The aforementioned phased approach to QMS implementation is *not* intended to imply a strictly sequential flow of phases and activities; it is intended only to emphasize a logical and structured process flow for implementing a QMS. Generally, several activities during the implementation will be executed in an iterative and overlapped fashion. Experience from use of a new process may result in revisions to it and its supporting documentation. This in turn may necessitate retraining of employees on the revised process, or at least that they be made aware of the changes by being provided revised QMS documentation if any. In fact, such evolution of an organization's processes is never-ending, and is a natural consequence of continuous process improvement efforts. However, the degree of process change is greater while the processes are relatively new. This is because the processes may need to be retooled frequently by the process owner in response to feedback from process practitioners. Processes take time to stabilize to where they are deemed satisfactory for the intended purpose, their performance is predictable, and their capabilities are fully determined (from experience). Once a process has been stabilized (which can take a few months), it can be improved incrementally. Such improvements are usually sufficient until factors such as a

* Note that, in the Gantt chart representation in Table 5, the continuous improvement phase is not included because the implementation of the QMS infrastructure is typically complete *prior* to start of this phase. Further, the continuous improvement phase is a never-ending phase and is therefore not appropriate for inclusion in a time-constrained Gantt chart.

change in the environment (e.g., organizational change), a change in requirements, or other such factors cause it to be reengineered or significantly revised.

Finally, bear in mind that the implementation schedule and plan established at this stage are preliminary, and will need to be refined as the implementation progresses and the QMS implementation team gains a better understanding of the underlying tasks. It is recommended that the initial baseline of the implementation schedule not be established until after the gap analysis in the QMS definition phase has been completed. As stated earlier, the gap analysis provides valuable insight into the current state of the organization. It therefore enables the QMS implementation team to better plan the project, taking into consideration the current situation. If there is sufficient forethought, and adequate contingency is included for each implementation activity (refer to Table 3), future refinements to the implementation schedule should result only in changes to some of the due dates for the implementation activities, without significant impact on the overall schedule.

Once the implementation gets under way and new information becomes available, the implementation schedule and plan should be revised when appropriate. At any time, the current situation or extenuating circumstances may necessitate a replanning of certain activities, or identification of new activities. The implementation schedule should reflect current realities and should be feasible, given current circumstances in the organization. It may also be used as a progress-tracking tool to reflect tasks completed and track the status of those in progress (by using the progress-tracking feature available in most off-the-shelf software used for preparing Gantt charts).

2.3 MECHANISMS TO MANAGE THE IMPLEMENTATION

A good project plan is not sufficient to ensure the success of the project. One also must establish control mechanisms to ensure project execution proceeds in accordance with the established plan. Therefore, the QMS implementation team must establish means to:

1. Track implementation progress
2. Plan for and control everyday implementation tasks
3. Plan for preventive actions needed to mitigate implementation risks
4. Plan for contingency in case known risks evolve into problems
5. Take timely corrective action when execution deviates from plans

In order to perform effective implementation tracking and oversight, the following simple tools have been found to be very effective.

Table 4 QMS Implementation Project Phases and Milestone Dates

Activity	*Responsible*	*Planned Start Date*	*Planned Completion Date*
Quality Management System Planning Phase		**July 2, 2003**	**Aug. 22, 2003**
Prepare QMS implementation plan	Management representative	July 2, 2003	Aug. 1, 2003
Establish QMS implementation team	Management representative	July 14, 2003	July 25, 2003
Define QMS implementation team charter	Management representative	July 21, 2003	July 25, 2003
Quality training for QMS implementation team (as needed)	Management representative	July 28, 2003	Aug. 22, 2003
QMS Planning Complete		Aug. 22, 2003	Aug. 22, 2003
Quality Management System Definition Phase		**Aug. 4, 2003**	**Feb. 27, 2004**
Perform preliminary gap analysis	Management representative	Aug. 4, 2003	Aug. 15, 2003
Revise implementation plan and schedule	Management representative	Aug. 18, 2003	Aug. 22, 2003
Correct critical quality discrepancies	QMS implementation team (and process owners and process practitioners, as needed)	Aug. 18, 2003	Sept. 19, 2003
Conduct high-level process mapping	QMS implementation team (and process owners and process practitioners, as needed)	Sept. 1, 2003	Sept. 19, 2003
Create process documentation for high-level processes (including verification and validation of the documentation)	QMS implementation team (and process owners and process practitioners, as needed)	Sept. 22, 2003	Oct. 31, 2003
Conduct lower-level process flowcharting	QMS implementation team (and process owners and process practitioners, as needed)	Sept. 22, 2003	Feb. 27, 2004

Create process documentation for low-level processes (including verification and validation of the documentation)	QMS implementation team (and process owners and process practitioners, as needed)	Sept. 22, 2003	Feb. 27, 2004
QMS Definition Complete		Feb. 27, 2004	Feb. 27, 2004
Quality Management System Refinement Phase		**March 1, 2004**	**May 14, 2004**
Verify complete QMS	Quality assurance department	March 1, 2004	March 31, 2004
Validate complete QMS against quality requirements (of the organization, and of a QMS standard, if used)	Quality assurance department	March 1, 2004	March 31, 2004
Perform necessary rework, re-approve, and publish final QMS documentation	QMS implementation team (and process owners and process practitioners, as needed)	April 1, 2004	May 14, 2004
QMS Refinement Complete		May 14, 2004	May 14, 2004
Quality Management System Deployment Phase		**May 17, 2004**	**Feb. 18, 2005**
Perform employee training on QMS	Management representative	May 17, 2004	July 9, 2004
Process establishment period	Not applicable	May 17, 2004	Oct. 29, 2004
Perform internal quality audits (final gap analysis)	Quality assurance department	Nov. 1, 2004	Dec. 21, 2004
Implement corrective and preventive actions (based on audit results)	Audited personnel (and management personnel)	Nov. 15, 2004	Jan. 28, 2005
Follow-up audit to verify implemented corrective and preventive actions	Quality assurance department	Jan. 31, 2005	Feb. 18, 2005
QMS Deployment Complete		Feb. 18, 2005	Feb. 18, 2005
Continuous Improvement Phase		**Feb. 21, 2005**	**Never-ending**
Use various mechanisms for continuous improvement, such as metrics, quantitative quality objectives, internal audits, customer satisfaction data, lessons learned from project postmortems, and so on.	All employees	Feb. 21, 2005	Never-ending

Table 5 Gantt Chart for QMS Implementation Schedule

ID	Task Name	Start	Finish
1	**QMS IMPLEMENTATION**	**Wed 7/2/03**	**Fri 2/18/05**
2	**QMS Planning Phase**	**Wed 7/2/03**	**Fri 8/22/03**
3	Prepare QMS Implementation Plan	Wed 7/2/03	Fri 8/1/03
4	Establish QMS implementation team	Mon 7/14/03	Fri 7/25/03
5	Define QMS implementation team charter	Mon 7/21/03	Fri 7/25/03
6	Quality training for QMS implementation team	Mon 7/28/03	Fri 8/22/03
7	QMS planning complete	Fri 8/22/03	Fri 8/22/03
8	**QMS Definition Phase**	**Mon 8/4/03**	**Fri 2/27/04**
9	Perform preliminary gap analysis	Mon 8/4/03	Fri 8/15/03
10	Revise implementation plan (and schedule)	Mon 8/18/03	Fri 8/22/03
11	Correct critical quality discrepancies	Mon 8/18/03	Fri 9/19/03
12	Conduct high-level process mapping	Mon 9/1/03	Fri 9/19/03
13	Document high-level processes	Mon 9/22/03	Fri 10/31/03
14	Conduct lower-level process flowcharting	Mon 9/22/03	Fri 2/27/04
15	Document low-level processes	Mon 9/22/03	Fri 2/27/04
16	QMS definition complete	Fri 2/27/04	Fri 2/27/04
17	**QMS Refinement Phase**	**Mon 3/1/04**	**Fri 5/14/04**
18	Verify complete QMS	Mon 3/1/04	Wed 3/31/04
19	Validate complete QMS	Mon 3/1/04	Wed 3/31/04
20	Final rework and reapproval	Thu 4/1/04	Fri 5/14/04
21	QMS refinement complete	Fri 5/14/04	Fri 5/14/04
22	**QMS Deployment Phase**	**Mon 5/17/04**	**Fri 2/18/05**
23	Perform employee training on QMS	Mon 5/17/04	Fri 7/9/04
24	Process establishment period	Mon 5/17/04	Fri 10/29/04
25	Perform internal quality audits	Mon 11/1/04	Tue 12/21/04
26	Implement corrective/preventive actions	Mon 11/15/04	Fri 1/28/05
27	Follow-up on internal quality audits	Mon 1/31/05	Fri 2/18/05
28	QMS deployment complete	Fri 2/18/05	Fri 2/18/05

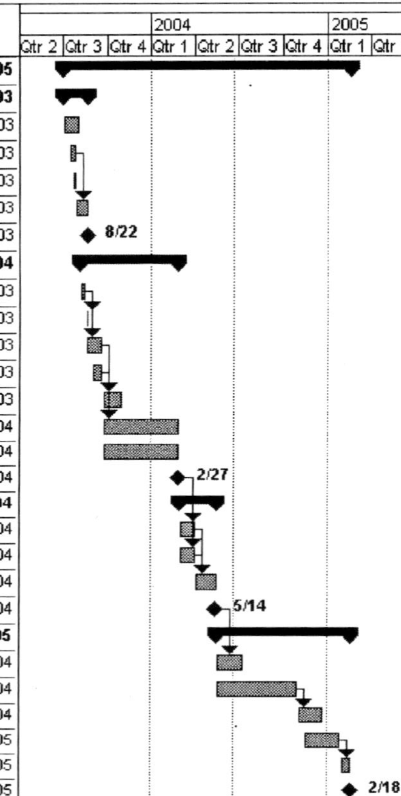

2.3.1 QMS Documents Status Sheet

A QMS implementation entails the creation of a number of QMS documents. The QMS implementation team must be able to track and report the status of each QMS document.* A simple table that lists at least the name of each QMS document, along with due date and status, serves as a valuable tool for monitoring implementation progress against plans. Such a table also enables the implementation team to provide status upon request by management personnel and other employees. It is recommended that such a status sheet be maintained by the quality assurance department, be updated on a daily basis (minimally on a weekly basis), and be stored in a location that is easily accessible to all employees (including senior management).

An example of a QMS documents status sheet is shown in Table 6. The sheet is divided into sections based on departments that own the respective QMS documents. This does not mean the owning department is the sole department to which the respective QMS document applies. As we know, the majority of the processes in organizations are cross-functional, and span more than one department. However, there is usually one department that executes a major part of a process, and therefore, a management person belonging to that department may be reasonably identified as the process owner. In other cases, when a process applies equally to various departments, one particular department generally is acknowledged to possess subject matter expertise or majority stake in the faithful execution of the process, and is assigned process ownership. In yet other cases, concerns about impartiality may cause the process ownership to be assigned to an independent department, such as the quality assurance department.

As an example, in Table 6, the "Document, data, and record management procedure" is listed under the documentation department, but this does not mean this is the only department expected to comply with the procedure. In fact, this procedure applies to all employees within the organization, but the organization might identify a management person belonging to the documentation or quality assurance department as the process (or document) owner. This is because these two departments typically are the most knowledgeable in all aspects of document management, and they have a major stake in the faithful execution of this process. While one could argue that the quality assurance department has a big stake in the correct execution of all processes, this argument should not

* Here, the reference is to QMS *infrastructure* documents, such as, procedures, work instructions, forms, templates, and so on (i.e., Levels 1 to 3 in Figure 6). Project documents and records (i.e., Level 4) pertain to a *specific project* and are not included here. Their status is tracked by the project manager for the specific project.

Table 6 QMS Documents Status Sheet

Last Updated: Oct. 28, 2003

#	*QMS Document Name*	*Owner*	*Author*	*Due Date*	*Completion Date*	*Status*
Contracts Department						
1	Contract specification and review procedure	Peter Oren	Peter Oren	Nov. 14, 2003		Not started
2	Contract review checklist	Myriam Najabi	Mike Salone	Nov. 14, 2003		Draft checklist available for initial review by quality assurance department
Documentation Department						
1	Document, data, and record management procedure	Mike Hollosi	Bill Lehto	Nov. 3, 2003	Oct. 28, 2003	Approved and released
Product Development Department						
1	Engineering lifecycle procedure	Rohinton Mistry	Al Longman	Dec. 3, 2003		Draft version approved by quality assurance department. Formal inspection meeting scheduled for Nov. 25, 2003.
2	Product design document template	Sharon Lu	Sharon Lu	Dec. 16, 2003		Not started

be used to assign ownership of all or the majority of the processes to the quality assurance department. Quality is every employee's responsibility, and not solely that of the quality assurance department. Therefore, departments must be made responsible for defining their processes, adhering to the defined processes, and for continuous improvement of those processes.

2.3.2 QMS Implementation Action Items Log

Each of the key activities in the QMS implementation phases shown in Table 4 entails the execution of a number of lower-level tasks. Many of these tasks need to be identified and planned for on a daily or weekly basis. This is because the need for most of these tasks can be recognized only during the implementation of the related higher-level activity. PMC members and process owners have the responsibility of identifying, coordinating, and tracking tasks related to their processes and departments. Similarly, the quality assurance department has the responsibility of identifying, coordinating, and tracking tasks that are the responsibility of the quality assurance department, or pertain to interdepartmental issues. Therefore, in addition to the QMS documents status sheet, the quality assurance department requires tools to:

- Control implementation progress (by using an action items log), and
- Manage ongoing implementation risks (by using a risk management plan).

The action items log should be used to perform day-to-day task management (see Table 7). It should contain a detailed list of tasks to be performed in each of the implementation phases. This log serves as a short-term work plan, with tasks being identified and planned for on a daily or weekly basis (including recording of tasks for future phases as the need for them is identified). Once an action item is completed, it may be retired from the log, or retained and marked as completed (and grayed out as shown in Table 7).

Quality assurance department personnel should collectively review the QMS documents status sheet, action items log, and risk management plan (explained in the next section) on a weekly basis at minimum. This enables the quality assurance department to exercise control over project execution, and facilitates timely identification and resolution of project risks and identification of schedule slippages. These slippages then can be discussed with appropriate PMC members to identify corrective action, or escalated to the periodic senior management review, if necessary.

The periodic review of the action items log should entail the following:

- Determination of whether the tasks planned for completion in the previous week have been satisfactorily completed;
- Planning for tasks to be executed in the following week; and
- Determination of additional tasks required in the current or subsequent implementation phases.

2.3.3 QMS Implementation Risk Management Plan

The QMS implementation risk management plan is used to record emerging implementation risks, and to perform real-time management of known risks. A risk management plan enables the QMS implementation team to minimize the likelihood of occurrence and impact of unexpected events or factors that can threaten project success. Generally, in projects, risks that need to be managed are related to scope, cost (resources), and time (schedule). Ongoing risk management includes determination of:

Probability of risk — What is the likelihood that a risk will materialize and evolve into a *real* problem? Probability of a risk may be categorized as high, medium, or low, each with a specific weight associated with it — say, 3, 2, and 1 respectively.

Severity of risk — The severity of risk is based on the anticipated consequences of the risk evolving into a problem. The worse the anticipated consequences, the greater the risk severity. Again, severity of risk also can be categorized as high, medium, or low, and weighted accordingly (as with probability of risk).

Risk exposure — This is the product of the probability of risk and severity of risk. The higher the two are, the higher the overall risk exposure.

Risk priority — Once the risk exposure for each risk has been computed, all the risks can be sorted in descending order of risk exposure, and numbered sequentially. This assigns a risk priority to each risk, beginning with risks that have the highest risk exposure (see Table 8).

Risk mitigation — This is the planned preventive action to mitigate a known risk and prevent it from evolving into a real problem. In essence, risk mitigation is *proactive* in nature.

Risk contingency — This is the planned corrective action to be executed in case a risk actually evolves into a problem. In essence, risk contingency is *reactive* in nature.

For each risk mitigation and contingency, a risk owner is also identified. The risk owner is responsible for monitoring each risk and executing the related action. Once a risk has been mitigated and no longer exists, it may be retired from the log, or retained and marked as completed (and grayed out as shown in Table 8).

Table 7 QMS Implementation Action Items Log

Last Updated: 7/16/03

#	*Implementation Activity*	*Assigned To*	*Due Date*	*Status*
Quality Management System Planning Phase				
1	Prepare high-level presentation of draft QMS implementation plan for presentation to senior management.	Ronen Spell	7/23/03	7/18/03: Presentation draft is ready. To be reviewed by QA on 7/21/03.
2	Set up meeting with Directors to present overall implementation approach and discuss possible candidates for PMC.	Ronen Spell	~~7/16/03~~ 7/18/03	Meeting scheduled for 7/18/03. Due date of task extended.
3	Prepare draft of QMS implementation team charter for review by VP of Quality.	Kistna Sanders	7/23/03	In progress
4	Investigate vendors and cost for Quality 101 and process mapping training for QMS implementation team.	Yana Erikson	~~7/14/03~~ 7/23/03	Quotes received from three vendors. Will be discussed at next QMS Steering Group meeting. Due date of task extended. See Risk 1 (Table 8).
5	Investigate cost of preliminary gap analysis by independent third party.	Yana Erikson	7/9/03	Done. Cost is $4500 for three-day assessment.
6	Talk to VP of Quality to decide if preliminary gap analysis should be performed in-house or by third party.	Kistna Sanders	7/14/03	Done. Gap analysis will be done in-house.
7	Perform preliminary investigation of how (and who) to perform in-house gap analysis.	Yana Erikson	7/22/03	

Quality Management System Definition Phase				
1	Provide each PMC member with relevant QMS standard requirements for implementation.[a]	Kistna Sanders	TBD	Future action
2	Train PMC members on process flowcharting and documenting QMS procedures	Yana Erikson	TBD	Future action
3	Create process template and e-mail it to all PMC members.	Kistna Sanders	TBD	Future action
	(Add additional tasks as they are identified)			
Quality Management System Refinement Phase				
	(Add tasks as they are identified)			
Quality Management System Deployment Phase				
	(Add tasks as they are identified)			
Continuous Improvement Phase				
	(Add tasks as they are identified)			

[a] Note that this action item is relevant to organizations that are implementing a QMS to conform to a particular quality management system standard.

Table 8 QMS Implementation Risk Management Plan

Last Updated: 7/16/03

Risk Priority	*Risk Description*	*Probability (High = 3; Med = 2; Low = 1)*	*Severity (High = 3; Med = 2; Low = 1)*	*Risk Exposure = Probability × Severity of Impact*	*Risk Mitigation*	*Risk Contingency*	*Notes*
1	Delay in quality training vendor selection decision by the QMS steering group may cause schedule delays in the QMS definition phase.	2	2	4	Action: VP of Quality will propose vendor by e-mail to QMS steering group for approval. Risk owner: Ronen Spell	Action: Request the selected vendor to prioritize Quality 101 course over process mapping course. Risk owner: Ronen Spell	See Action Item 4 in action items log.
2	Senior management may not approve current phased approach to QMS implementation, wherein certain areas are proposed to be postponed to a future phase in order to meet the completion time frame of Q1 2005.	2	3	6	Action: Explain to senior management why a phased approach is being adopted, and provide a high-level plan for subsequent QMS implementation in areas proposed to be postponed. Risk owner: Ronen Spell	Action: Prepare an alternative plan wherein all areas are in scope of implementation, but definition of lower-level processes is postponed to a future phase. Risk owner: Ronen Spell	Closed. Senior management agreed to postpone certain areas to a future phase.
3	And so on						

2.4 COMMUNICATION *

Communication is a key element of any QMS implementation and should never be overlooked. This includes communication not only within the QMS implementation team, but also up, down, and across the organization. The project manager and QMS implementation team also must communicate with senior management and employees across the organization (stakeholders).

Good and constant communication serves various objectives:

- It enables the QMS implementation team to articulate its vision, implementation approach, and anticipated benefits to all personnel. This promotes a shared vision and common understanding of the implementation objectives and means for achievement of those objectives. This in turn promotes consistency in approach and closer alignment of all participants.
- It helps keep the organization informed about implementation progress and forthcoming developments. For example, employees should be informed about the key elements of the QMS that have been established (and require compliance).
- It helps elicit employee feedback on the implementation approach and benefits realized.
- It helps sustain momentum and interest within the organization towards the QMS implementation, and fosters a strong sense of participation among all employees.
- It enables the QMS implementation team to draw the attention of employees to certain aspects of the QMS (as needed).
- It promotes constant involvement of senior management by keeping them informed about the progress of the various departments in the implementation of and adherence to the QMS.
- It serves as a vehicle for acknowledging excellent work or progress by departments or individuals in implementing the QMS.

Perhaps most importantly, the increased visibility of the quality effort due to effective communication helps promote a new approach to business — one centered on quality management and continuous improvement. However, caution should be exercised to ensure that the QMS has sound underpinnings in the operational processes of the company, and is not merely limited to impressive slogans and posters throughout the organization.

During QMS implementation and beyond, the following vehicles for communication can be used for communication between QMS implementation team members, employees, and senior management.

* Parts of this section are reproduced with permission from Nanda, Vivek (Vic), *ISO 9001:2000 — Achieving Compliance and Continuous Improvment in Software Development Companies*, ASQ Quality Press, 2003.

2.4.1 Communication with QMS Implementation Team Members

PMC members are a vital interface between the quality assurance department and various departments in an organization. Therefore, communication with the QMS implementation team members enables the project manager and quality assurance department personnel (the core team of the implementation, as explained earlier in this chapter) to propagate their message and plans to the respective departments via the appropriate PMC members and site coordinators. Solicitation of feedback from the PMC members by the core team helps encourage their active participation, and promotes a feeling of collective responsibility and joint ownership of the system — factors that are vital for the ultimate acceptance and use of the QMS.

For the aforementioned reasons, the project manager should organize and chair periodic meetings of the QMS implementation team. Generally, participation of the site coordinators and PMC members is sufficient at these meetings, while other implementation team members might be invited as needed. Such meetings serve as a valuable forum for the free exchange of ideas between the core team and QMS implementation team members, and for involving QMS implementation team members in decision making and planning of next steps. The objectives that such meetings serve include the following:

- Review of overall implementation progress against plans
- Monitoring and control of project execution — that is, they facilitate timely corrective and preventive actions when progress deviates from plans.
- Planning of future activities and tasks
- Identification of new project risks, and management of existing risks

Records of such meetings, in the form of meeting minutes, should be maintained. The QMS implementation action items log and risk management plan should be updated to reflect the output of such meetings.

Additionally, appropriate QMS implementation team members (primarily PMC members) and relevant employees should review QMS documents prior to approval and release of the subject documents. This provides an opportunity for the QMS implementation team members to participate in review and approval of QMS documents that pertain to their processes, or to invite relevant personnel from their departments to participate in the reviews. This helps secure the buy-in of the various stakeholders that is necessary for the acceptance and use of the QMS documents.

2.4.2 Communication with Employees

In addition to communication meetings with the QMS implementation team members, it is important for the success of the QMS implementation that senior management and the quality assurance department communicate with all employees. For this purpose, senior management should emphasize the importance of quality, continuous improvement, and customer satisfaction at "all-hands" and staff meetings.

For the purpose of employee communication, strategically located bulletin boards and posters can be very useful. Bulletin boards can be used to display information that is pertinent to the QMS implementation phase in progress. For example, during the QMS planning phase, information such as names of PMC members from each department (to promote their awareness) and overall implementation process may be displayed, while during the definition phase, the displayed information may include a list of the QMS documents that have been created in each department. Care should be taken to keep the bulletin boards and posters brief yet sufficiently attractive, so that a passer-by would be tempted to stop and read the material. Displaying excess information is counterproductive and likely to discourage the employee from reading the posted information.

In addition to the above, periodic newsletters (preferably sent by e-mail) can be an effective means of communicating with all employees. Such newsletters should be brief, a page or less, and not sent so frequently that they would cause the employee to regard them as "junk mail" and discard them without reading. Depending on the overall time frame for implementation, a newsletter every two months usually is sufficient. The newsletter should draw the attention of the employees to the accomplishments in the QMS implementation to date, discuss what to expect in the coming days (until the next newsletter), and recognize employees, including PMC members, who have made significant contributions in the development of the QMS.

Another vehicle that may be used for internal communication is a quality Web site on the organizational intranet. During QMS implementation, this Web site can be used to communicate departmental progress, including accomplishments and current and planned activities. The Web site can be used for soliciting improvement suggestions from employees and for providing information regarding corrective and preventive actions in implementation. The Web site also can be used to share process and product measurement information with employees (access can be restricted if required). Note that such a Web site is meant to be primarily for informational purposes only, and is separate from the controlled repository of QMS documentation that also can be implemented electronically.

Other useful means of communicating with employees include sending out "quality alert" e-mails to appropriate departments regarding pertinent quality issues — for example, to clear up misunderstandings regarding key elements of the QMS, to caution against observed discrepancies from the QMS, or to draw attention to key requirements of the QMS. Quick reference cards also might be distributed to succinctly summarize key elements of the QMS. As an example, the sample quick reference card shown in Figure 9 lists key elements of the QMS, such as:

- Corporate quality policy
- Structure of the QMS
- Key procedures, work instructions, form, and templates
- Internal quality training courses
- Name of the organization's management representative

In order to tailor the QMS information as per the unique needs of each department, such quick reference cards may be produced on a "departmental" basis. It is preferable to use an attractive color background so that the quick reference cards are easily noticeable at all times. A few copies of the quick reference cards must always be made available at locations that are frequently visited by the employees, such as the company cafeteria.

ABC CORPORATION

Quality policy

"To delivery superior quality products and services to all customers by measurement driven continuous improvement"

Our QMS structure:

***Level* 1:** High-level QMS documents, e.g., Quality Manual

***Level* 2:** Procedures

***Level* 3:** Work instructions, forms, templates, checklists, etc.

***Level* 4:** Project documents and records

Fold Here

Key procedures:

- Product life cycle procedure
- Project planning procedure
- Document, data, and record management procedure
- Product test procedure
- *<list additional docs>*

Key work instructions, forms, and templates:

- Product requirements template
- System test case template
- Inspection record form
- *<list additional docs>*

Fold Here

Quality training courses:

- QMS-100: QMS overview training
- QMS-200: Product life cycle training
- QMS-210: Project Management processes training
- QMS-300: Product testing techniques training
- *<list additional trainings>*

Questions? Contact the quality management representative:

Kevin Knight
kk@abc.com
Ph: Ext. 2752

Figure 9 Sample QMS Quick Reference Card

Quality contests and associated prizes also serve as an effective way to encourage employee participation in the QMS establishment effort. Such contests help familiarize the employees with the QMS terminology and promote use of the QMS. For example, employees may be quizzed on elements of the organization's QMS.

Finally, bear in mind that communication with employees should not be one-way (from QMS implementation team members to employees), and employees should be encouraged to provide their input for QMS implementation and improvement as well. For example, employees might provide their input via their PMC member, or by contacting an appropriate member of the QMS implementation team. For medium-sized or large organizations, it is preferable that a formal mechanism be available to the employees to submit their suggestions and feedback (including anonymously). Examples of mechanisms that can be used to collect employee input include the company intranet, e-mail, and employee suggestion box. An example of a procedure for handling process improvement suggestions is provided in Appendix C.2.

2.4.3 Communication with Senior Management

As stated earlier, management commitment and continued management support are essential for the long-term success of an organization's QMS. During QMS implementation, it is important that the quality assurance department keep senior management informed regarding the implementation progress in all the departments. For this purpose, a scale and criteria to measure progress should be defined, and the methods used to measure progress should be communicated to all PMC members. Sample scale and criteria are shown in Table 9.

PMC members should be made aware that detailed QMS implementation progress status is reported at management reviews. Such status reporting to senior management helps ensure that all departments accord high priority to their respective tasks. It is useful from a senior management perspective to be provided an overall slide showing quantitative progress of each department (see Figure 10), along with one slide, per department, to report specific issues (see Figure 11). As an example, Figure 10 shows a high-level overview of the status of QMS implementation in a company. All departments are awarded a score on a scale of 0–100 (using the scale in Table 9). The figure shows the score for each department for the current month, and for the previous month. (The difference between the two is the progress made by the department since the last month.) As an example, the detailed status of a department, and the breakdown of its score for the current month, are shown in Figure 11.

Table 9 Scale and Criteria to Measure QMS Implementation Progress

Scale	*Points*	*Explanation*
0–20	20	Draft versions of all requested QMS documents have been produced and are ready for informal review by quality assurance department. For example, if four QMS documents are requested from the department, you should consider each draft document to be worth 5 points.
21–40	20	All draft documents were reworked after informal review, and are ready for interdepartmental formal review.
41–60	20	All documents have been reworked and approved after formal review.
61–75	15	All employees in the department have been trained on applicable QMS documents.
76–90	15	The department has been audited by the quality assurance department and has responded with an acceptable corrective action plan.
91–100	10	The quality assurance department has performed a corrective action follow-up audit, and QMS implementation is now complete.

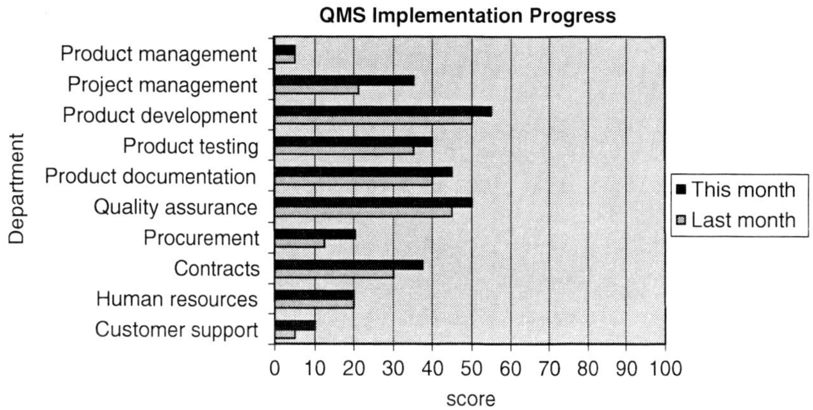

Figure 10 Overall QMS Implementation Progress of All Departments in an Organization

QMS Implementation Status

■ **Department: Project Management**

- PMC member: Marie Smith
- Current score: 35%
- Progress (since last month): 14%
- Summary: Three documents needed for QMS implementation:
  - Project Plan Template: Was formally reviewed, reworked, and approved for release
  - Patch Planning procedure: Was informally reviewed by quality department, rework also complete
  - Project Tracking and Control procedure: Draft version has been delayed by another 2 weeks
- Score:(14/20+14/20+7/20+0/15+0/15+0/10)

Figure 11 Project Management Department — QMS Implementation Status Update

The quality assurance department should not spend undue time trying to arrive at the most accurate score for a department. This is not intended to be an exact science, but to serve as a high-level yet reasonably accurate estimate of the progress of each department. The quality assurance department will need to exercise discretion while awarding points to account for progress from the last progress-reporting period. The importance of such progress reporting with measurement data, though sometimes cumbersome to compute, should not be overlooked. Care should be taken to emphasize to senior management and personnel that progress percentages of their respective departments are systematically derived and are not arbitrary, yet are not intended to be accurate to the last digit. Therefore, it is important for personnel to keep this in perspective and not argue each individual score awarded by the quality assurance department.

2.4.4 External Communication and Networking

All the communication mechanisms described so far are internal communications; i.e., they are meant to be used within the organization. If an organization were implementing a QMS from scratch, it would be interested in knowing how other companies in its business domain implemented their QMS. Participation in a quality network, such as a local chapter of the American Society for Quality (or other national quality association), can be very helpful for this purpose. Guidebooks such as this one can be a useful source of information, but it is always advantageous to network with other companies and benefit from their experiences.

Some of the questions that you would want to ask other companies during such "external communications" include:

- What is the scope of your QMS?
- What is the size of your organization, and how long did it take you to implement a QMS?
- What high-level process and strategy did you follow to implement the QMS?
- What was the state of the system at the start of QMS implementation? For example, was some QMS documentation available and already in use?
- Was there a dedicated effort (project) to implement the QMS within a specific time frame? If so, would you be willing to provide a copy of your project plan?
- How much did the QMS implementation cost?
- How long did it take to implement the QMS?
- What is the structure of their QMS?
- What were the lessons learned?

By drawing upon the experiences of other companies, an organization can attempt to replicate other companies' successes by adopting and applying recommended practices (with some tailoring) to its unique situation. Similarly, it aids awareness of implementation pitfalls that other companies caution against, and an organization can plan to avoid them. Finally, such networking in the industry also helps publicize an organization's quality improvement effort and gain the attention of potential customers.

2.5 MOTIVATING AND RECOGNIZING EMPLOYEES

A QMS implementation requires time and commitment from all employees. Many employees will participate in the implementation effort by creating required QMS documents. Outstanding employee efforts should be recognized by the quality assurance department to thank the employees for their contribution and to ensure their continued support. Employee recognition can be made by means of the periodic newsletter described earlier. Individual contributors also can be recognized at company events or meetings by requesting senior management to hand out awards, such as employee recognition certificates along with gift certificates. Such recognition not only would be appreciated by the employees who have devoted significant amounts of their time to accomplish the requested tasks, but also would serve as an incentive for other employees to participate in the QMS establishment. The quality assurance department also should celebrate milestone achievements with the QMS implementation team members by organizing team lunches on such occasions, or by other appropriate means.

3

QMS DOCUMENTATION PLANNING

Perhaps the most significant undertaking in a QMS implementation is the creation, rollout, and use of documentation across an organization. An organization's QMS documentation comprises documentation at four different levels of abstraction (level 1 to level 4) as explained in Chapter 2 (see Figure 6).

Before an organization begins creating QMS documentation, it must plan for it. Planning for QMS documentation refers to planning for "infrastructure-level" QMS documentation only — documentation at levels 1, 2, and 3 in Figure 6. QMS documentation at level 4 comprises "evidence level" documentation — documentation that provides evidence of use of the QMS, such as project documentation. Planning for such documentation is performed during the planning phase for a specific project, and therefore is outside the purpose and scope of this chapter.

Planning for QMS documentation should address the following three elements:

- Overall strategy for creating QMS documentation;
- Documentation management and control mechanisms; and
- Process for creating QMS documents.

Agreeing upon such issues up front will facilitate the creation of the QMS documentation. Document creation can proceed unhindered once the necessary guidelines are in place to support the creation, review, approval, and dissemination of QMS documentation. Each of the aforementioned elements of QMS documentation planning is briefly described next. (Detailed explanations are included in the rest of the chapter.)

Documentation strategy — This is perhaps the most critical element of QMS documentation planning. A well-thought-out, rational approach to documenting your QMS will enable you to rapidly develop a QMS that works and has sufficient but not excessive detail, and, perhaps most importantly, will provide you with a documentation set that is usable, not one written primarily to appease external quality auditors. After all, the primary users of an organization's QMS documentation are its employees, not outside parties. Brainstorming the documentation strategy entails obtaining answers to the following questions:

- What approach should be adopted for documenting the QMS (top-down or bottom-up)?
- Up to what level of detail should processes be documented (breadth and depth of documentation)?
- How can the QMS documentation be kept relatively stable and immune from minor changes in the organization or its processes?

Documentation management and control — Documentation management and control are a key element of an organization's QMS, providing mechanisms to ensure that documentation in the organization is uniquely identifiable, reviewed and approved by the appropriate authority prior to release, made available to users, kept current, changed in a controlled manner, and archived when obsolete (to prevent unintentional use). This entails answering questions such as:

- What types of QMS documents are required?
- How should the QMS documents be logically structured?
- How should the QMS documents be uniquely identified?
- Who should review and approve documents?
- How should changes to QMS documents be identified and controlled?
- How should superseded (or obsolete) documents be handled?
- How should documents of external origin be controlled?
- Should the QMS documentation be stored in a physical (i.e., hard-copy) repository or an electronic (i.e., virtual) repository?
- How should employees be provided access to controlled QMS documentation; that is, how should the QMS documentation repository be published?
- How should the published QMS documentation be organized to maximize ease of use for employees?*

* This issue is closely related to the issue of logical structuring of the QMS documentation. The actual physical structuring of QMS documentation (shown in Figure 14) should be very similar to the logical structuring of the QMS (shown in Figure 6).

Documentation process — The third element of QMS documentation planning entails the establishment of a process for the creation, review, rework, approval, and final release of QMS documents.

The above elements are discussed in the following subsections.

3.1 DOCUMENTATION STRATEGY

The top-down approach to implementing and documenting the QMS is highly recommended (as opposed to the bottom-up approach). (Refer to the previous discussion on implementation strategy in Chapter 2.) For example, the overall process that results in a delivered product — the product development process — may be documented in a product development procedure and/or in a product development process map. Within the product development process, the process that results in a formalized set of product requirements may be identified as the product requirements definition subprocess that is documented in a product requirements definition procedure. Similarly, the process that results in a formal documented design for the product may be identified as the product design subprocess that is documented in a product design procedure. The procedures describing the subprocesses and their interaction should be supported by additional QMS documentation, as appropriate. For example, the product design subprocess should be supported by QMS documentation such as a design document template, a product design guidelines document, and other needed documentation.

Another important issue that needs to be addressed regards level of detail. What is the right level of detail to include in the documentation, so that it enables correct and consistent process execution, and minimizes impact (on QMS documentation) of minor changes in the business processes and organizational structure?

Any breakdown or inconsistency in process execution does not necessarily result from insufficient QMS documentation, but might result from unclear or ambiguous QMS documentation, inadequate employee training, or other factors. Therefore, organizations should be careful in how they respond when process execution deviates from requirements. Creating more QMS documentation is not necessarily the right solution. Sometimes, review of existing QMS documentation to identify and correct deficiencies, or conducting employee training to emphasize key aspects of a process, is the preferred solution.

Below are some guidelines to follow to ensure QMS documentation has the right amount of detail:

- Include all information that is specifically required to be documented as per the applicable quality management system standard.

- Include only as much information as is necessary to ensure effective planning, operation, and control of processes. Other factors that have a bearing on extent of QMS documentation include size of the organization, type of activities, complexity of the process being documented, and competency level of employees executing the processes.
- QMS documents should be written so that they need minimum change, if any, for minor operational or organizational changes. Some useful tips to accomplish this are:
  - Always refer to roles (or functional area) that are involved in the execution of a process, as opposed to referring to individuals by name. For example, instead of saying that Mark Peterson produces the product requirements document, state that the requirements document is produced by the requirements engineer.
  - Often departments in organizations are renamed or merged into other departments. Such an organizational change can have a huge impact on the QMS documents. To minimize the impact of such reorganizations, instead of referring to departments by name, refer to them by the "function performed." For example, if your organization has a product test department that is called "system test department," then refer to this department as the test department. If this department subsequently is renamed the "independent verification and validation department," you need not revise your QMS documents if the functional responsibilities for testing the product still reside with this group.
  - Do not document details of an activity in more than one QMS document. If an activity is described in a procedure X, and another procedure Y needs to refer to the same activity, state in procedure Y that detailed description regarding that activity can be found in procedure X. In the event of any change in the execution in that activity, it is far more convenient to revise one QMS document than to revise multiple documents.
  - When referring to another department's QMS documents, refer only to its high-level QMS documents, such as procedures. Avoid references to another department's work instructions or similar implementation-level documents. Implementation-level QMS documents typically are more prone to changes in content, title, and scope than are high-level documents. For example, department Y's procedure refers to department X's work instructions. Certainly it is quite inconvenient for department Y to revise its procedure when department X revises its work instructions such that the reference from department Y's procedure becomes inaccurate.

3.2 DOCUMENTATION MANAGEMENT AND CONTROL

3.2.1 Role of the Document Controller

Before discussing different elements of document management and control, it is necessary to introduce the concepts of *controlled documents* and *document controller*. A controlled document is one that is formally approved and is under formal version control. The different types of QMS documents discussed later in this section all are examples of controlled documents. Document controllers are people who coordinate, monitor, and enforce an organization's documentation management and control function. Depending on the size of an organization and/or its number of locations, the document controller function may be centralized or distributed. Generally, it is best to control documents that affect an entire company (e.g., company quality manual and operating procedures) in a centralized location. However, in case of companies with more than one location (or very large companies in a single location), QMS documentation that pertains to a specific location (or a specific function) may be controlled locally.

A document controller's responsibilities generally include, but are not limited to:

1. Verifying that documents submitted for storage and publishing are:
   a. In the correct format (that is, they adhere to standardized templates when applicable)
   b. Duly approved
   c. Accompanied by review records (when required)
2. Notifying appropriate personnel, such as the document author and management personnel from the affected area, when errors or discrepancies are observed.
3. Correctly storing and publishing (or distributing) controlled documents. This includes withdrawing copies of obsolete controlled documents.
4. Verifying that the documents are correctly numbered. When document numbers are issued manually for new documents, this task generally is performed by the document controller. Alternatively, a document number may be generated automatically.
5. Verifying that the documents are correctly versioned. In case of changes to previously approved documents, the document controller should verify that the document version accurately reflects the magnitude of change in the latest version (refer to explanation in Section 3.2.4).

6. Verifying that changes made to previously approved documents were properly authorized (that is, for revised documents, an approved document change request should be available)
7. Verifying that all changes made to previously approved documents are clearly identified. This is necessary because the document author may make document changes in addition to those that were authorized.
8. Notifying affected personnel in the event of a change to a previously approved document (or release of a new document)
9. Ensuring that all controlled documents are stored in a secure location
10. Maintaining a master list of controlled documents
11. Controlling documents of external origin. This includes clearly identifying documents of external origin and storing them in a secure location.
12. Authorizing internal documents for external release after verifying that approvals for the release have been obtained from relevant management personnel.

3.2.2 Types of QMS Documents

Quality Manual — A quality manual is the highest-level QMS document and is intended primarily to provide an overview of an organization's QMS. In case of medium and large product development companies, it is preferable to exclude details regarding the organization's processes from the quality manual. Such details should be embedded in the appropriate QMS documents, which must be referenced, as needed, from the quality manual. However, in the case of smaller companies, it may be appropriate to include the procedures in the quality manual itself.

An organization's quality manual is an invaluable document for its employees, customers (and potential customers), and other parties (such as third-party auditors). It therefore should reflect the organization's commitment to quality (in other words, answer *why* the organization is implementing a QMS) and describe how the organization ensures quality in its daily operations. Senior management should realize that it is responsible for the manual's content. This can be demonstrated by senior management approval on the quality manual. The quality manual must reflect the QMS accurately and be kept current at all times.

Typically, organizations structure their Quality Manual in one of two ways:

Standard-based quality manual — Most organizations that are implementing a QMS in accordance with the requirements in a particular quality management system standard, such as ISO 9001:2000, prefer to structure their manual to mimic the structure of the applicable quality management system standard. Such a quality manual includes separate sections (or subsections) for each of the requirements sections (or subsections) in the quality management system standard. Each section (or subsection) in the quality manual describes how requirements in the corresponding section (or subsection) in the quality management system standard are adhered to in the organization. When appropriate, the quality manual references relevant QMS documentation in explaining adherence to each quality management system standard requirement (or set of requirements).

Process-based quality manual — This structure is being used increasingly in organizations that have successfully transitioned to taking a process-oriented view of their QMS. Such quality manuals follow a top-down approach to describing the organization's QMS. The QMS is explained in the context of the organizational business processes, and not in the context of a quality management system standard (if one is in use*). This approach entails describing the high-level product development process map of the organization, followed by separate sections briefly describing each key process in the QMS. This includes describing the purpose and scope of each process, along with a reference to related QMS documentation.

A significant advantage of this structure is that it is not alien to employees; employees typically relate more closely to organizational processes than to the requirements in a quality management system standard. Therefore, employees invariably prefer such a quality manual, and consequently it gains wider acceptance for daily operations. A sample outline of such a quality manual is provided in Appendix B.

Procedure — A procedure is a documented high-level description of a process. Procedures constitute the first level of documentation below the quality manual. They serve as critical reference documents for anyone interested in knowing what a process entails. Procedures are not intended to provide the *how to* implementation details regarding a process. They describe:

---

* For such organizations, one of the significant differences between this type of quality manual and a quality management system standard-based quality manual is that it does not specifically describe how each requirement of the applicable quality management system standard is adhered to. (That exercise is performed by the external auditor during the registration process.) However, a traceability matrix between the quality management system standard and QMS documentation generally is included to guide the external auditor to the appropriate QMS documentation and records that demonstrate compliance to each requirement.

- *What* activities comprise a process;
- *When* each activity in a process is performed;
- *Who* performs the activities (roles and responsibilities); and
- *Where* (department and/or location) the activities are performed (refer to procedure template in Appendix D.1).

The decision to document a process in a procedure is made by the respective process owner in consultation* with the quality assurance department. (Refer to the next section for criteria that should be used to identify the need for a procedure.)

Procedures are usually interdepartmental, because organizational processes typically span multiple departments. Due to the interdepartmental nature of procedures, they should undergo cross-functional review by all departments involved in the process (or areas affected by the process) being documented. This helps ensure that the procedure accurately reflects the process and the interaction between various departments. As a general rule of thumb, a procedure should not be longer than three pages. (This refers only to the core content that describes the activities in the process; i.e., Section 5 in Appendix D.1.) If a longer procedure is needed, it is a good candidate for splitting into separate procedures.

Procedures are useful for communicating process information at all levels of management between departments. They also serve as a valuable starting point for training process practitioners. Because a procedure is intended to contain relatively high-level information regarding a process, practitioners typically need additional process documents, called work instructions, to execute their tasks.

Work instruction — A work instruction is a documented low-level description of a process. Work instructions describe *how* activities in a process are executed, and they constitute the first level of documentation below procedures. They provide a step-by-step description of tasks to be executed in order to accomplish each activity in the process. Work instructions typically are intradepartmental and are intended primarily for use by process practitioners. Due to the intradepartmental nature of work instructions, they should be documented and jointly reviewed by practitioners involved in executing the tasks documented. The practitioners have firsthand experience performing the tasks and therefore typically are the most knowledgeable, competent, and experienced personnel for

---

* By exercising control over what QMS documentation is created, the quality assurance department can ensure that proliferation of unnecessary QMS documentation is avoided, and that only required QMS documentation is created. Otherwise, the organization risks having an overabundance of QMS documentation that is impossible to maintain, comply with, and enforce.

providing information on the execution of specific tasks. When process practitioners do not have the requisite training or are otherwise unskilled for creating effective process documentation, this task may be performed by another appropriate person, such as the PMC representative for the department. In such a case, in order to secure buy-in of the practitioners and to ensure that the documentation accurately reflects practice, the work instructions should be documented with direct input and active involvement of the process practitioners. As a general rule of thumb, the core description of tasks in a work instruction should be limited to about four or five pages in length.

Not every procedure needs to be supported by underlying work instructions. Work instructions should be created on an as-needed basis when a need exists to provide detailed step-by-step guidance for process execution, to minimize variation, and to ensure consistency in process execution. In many cases, processes that are relatively straightforward and without inherent complexity (or sophistication) can be described adequately in a well-documented procedure such that competent personnel can faithfully execute them without compromising quality of process output. The decision to create a work instruction is made by the line manager or process owner who is responsible for the tasks.

The need to document a procedure or work instruction may be determined by using criteria such as:

- *Complexity* — Is the process or the activities in it sufficiently complex that that it needs to be supported by a documented procedure or work instruction? Or, is there need to elaborate and provide further explanation on a process documented in a procedure by creating a work instruction (e.g., the sequence of tasks to be performed during engine assembly for a passenger vehicle)?
- *Need for consistency* — Are there expectations regarding a high degree of discipline and consistency in executing a set of tasks (e.g., the sequence of inspections to be performed before final approval for release of electronic wire assemblies)?
- *Competence of personnel* — Is the competency level of the personnel executing the process such that it needs to be augmented with appropriate documentation to ensure the process is correctly executed? (For example, when new personnel or personnel with varying levels of competence are executing a process, work instructions can help ensure consistency and minimize errors.)
- *Size of organization* — Is the process executed by several personnel and/or in multiple locations? The greater the number of personnel involved in a process, or the greater the number of locations

at which a process is executed, the greater the likelihood of inconsistency in process execution.

- *Past problems* — Have there been instances in the past where inconsistencies in employees have been observed in process execution? In such cases, documenting an agreed way of executing a process can help.

The relationship between procedures and work instructions is depicted in Figure 12. Note that the figure depicts a relatively straightforward scenario where a process map consists of processes, which in turn consist of activities. As discussed earlier in this chapter, processes themselves may consist of subprocesses, especially in relatively large and complex organizations. In any case, the documentation rules remain the same — processes (and subprocesses) are documented in procedures and activities are documented in work instructions.

Templates and forms — A template is a skeleton for a document intended to be populated with specific information from use. Templates serve as guides for communicating the expected structure and content of a document. They help in ensuring consistency of format and content within a particular type of document, such as procedures or work instructions. For example, for documenting procedures, it is strongly recommended that a procedure template first be established. (See Appendix D.1.) The procedure author would start with the procedure template and populate it with information for the process being documented.

A form is used to record information, directly in the fields provided. Examples of forms are provided in Appendix D.

It is recommended that forms and templates have brief instructions embedded* in them to guide the user regarding the expected content in each section of the document.

The decision to create a template or form for use across departments should be made by the respective process owner in consultation with the quality assurance department (for reasons described earlier), while the decision to create a template or form for departmental use may be made by the respective line manager.

Project documents and records — Project** documents and records constitute the objective evidence to demonstrate that the QMS is being

---

* The *hidden text* feature in word processing software can be used for embedding such instructions as hidden text in templates and forms. Subsequently, once the document has been created, the instructions can be hidden by activating this feature.

** For organizations that do not execute their daily business operations as "project" activities, such as service delivery organizations, a project document or record may be regarded as equivalent to the usual business documents or records that the organization produces during daily business operations.

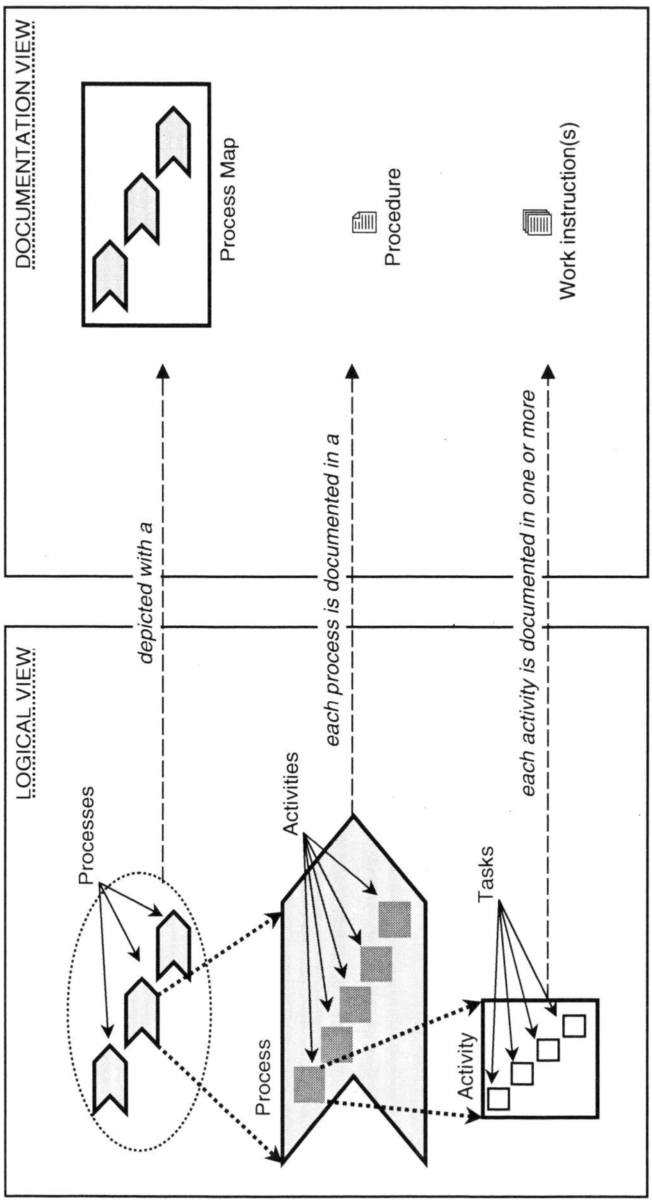

Figure 12 Relationship between process documentation and processes, activities, and tasks

used. Once a template or form is populated with project-specific information, it becomes a project document or a record. Therefore, project documents and records are types of QMS documents. For example, after a formal document inspection, once the inspection form has been duly completed, it becomes an "inspection record." Similarly, once the project plan template has been populated by the project manager, it becomes the project plan for a specific project. A project document or record thus constitutes a single instance of the applicable form or template, and serves as objective evidence that the QMS is being used. Because project documents and records demonstrate conformance (or lack thereof) with the QMS, they are closely scrutinized during quality audits as part of the objective evidence examination.

In addition to the aforementioned document types, additional types of documents may exist, including checklists, guideline documents, and workmanship standards. Such documents fall in the domain of "how to" documents, and can be used either in addition to or in lieu of work instructions (as appropriate). When such documents are for interdepartmental use, the respective process owner makes the decision regarding creation of these documents in consultation with the quality assurance department; when such documents are for intradepartmental use, the respective line manager makes the decision.

3.2.3 Document Numbering

In order to uniquely identify and control each QMS document, a unique document number and version should be assigned to each document. The organization should devise a document numbering convention that meets its needs. The numbering convention should be published in a QMS document, and should be enforced by the personnel in charge of (or tool used for) issuing the document numbers. An example of a document numbering convention is shown in Table 10. Examples of document numbers created by following the convention in Table 10 are shown in Table 11.

3.2.4 Document Versioning

Documents, due to their very nature, evolve. Once released, either for review or for use (after approval), a document typically will undergo revisions until it is withdrawn from use (or considered obsolete, but retained for archival purposes). Any change or set of changes made to a document since its last release necessitates that the new revision level of the document be formally identified. This is because the information contained in the new revision of the document is to some extent *different*

Table 10 Example of a Document Numbering Convention

	Document number format: AAA-BB-CCCC-DDDD
where,	
AAA:	Three-character identifier for department that is the document originator For example, "ENG" may denote engineering, "MKT" may denote marketing, "EXT" may denote the document is of external origin, and so on.
BB	Two-character identifier for document type For example, "PR" may denote procedure, "TP" may denote template, "CH" may denote checklist, and so on.
CCCC	Four-character alphanumeric identifier for project, product, or task
DDDD	Four-digit sequential number between 0001 and 9999. This sequential number uniquely identifies a specific document (irrespective of its version).

Table 11 Sample Document Numbers (based on numbering convention in Table 10)

AAA	*BB*	*CCCC*	*DDDD*	*Explanation*
ENG	PR	QUAL	0001	ENG-PR-QUAL-0001 is a procedure (PR) belonging to the QMS documentation set (QUAL). It was created by the engineering (ENG) department and it is the first (0001) controlled document created in the organization.
HRM	PL	ES04	0093	HRM-PL-ES04-0093 is a corrective action plan (PL) created by the human resources management department (HRM) to address the employee satisfaction survey findings for the year 2004 (ES04 project), and it is the 93rd (0093) controlled document created in the organization.

from that contained in the previous revision. Therefore, one must not only know what document to use, but also what revision level of that document to use. A document number, although sufficient to identify a document, is not sufficient to identify the revision level of the document. When a document version level is used together with a document number, it uniquely identifies the subject document *and* its revision level.

If the current approved version (version 1.0) of a document, identified by the document number TST-PL-VN04-1923 (which we can refer to as the *child document*), is based on information specified in a higher-level document REQ-SP-VN04-1901 (which we can refer to as the *parent document*), it is unclear what version of the parent document was used. This is important to know because the child document might be based on information specified in an obsolete version of the parent document that since has been changed. It is important to ensure that the child document is based on the *correct* version of the parent document, which can be done only by identifying the specific version of the parent document used.

There are various document versioning conventions in use in the industry. However, the most widely used versioning convention is as follows:

3.2.4.1 Versioning of Draft Documents

During the document creation process, a document evolves through a series of *draft* versions before it is finalized and formally approved. Such draft versions of a document can be identified in one of many different ways, such as:

- Version a, version b, version c, and so on; or
- Draft 1, draft 2, draft 3, and so on.

3.2.4.2 Versioning of the First Approved Release of a Document

Once a document has been formally approved for the first time, it should be identified by a standard version, for example, "Version 1.0." This approach fosters standardization within the organization regarding versioning of approved documents, and for identifying subsequent changes to them (as described below).

3.2.4.3 Versioning of Changes to Approved Documents

If a formally approved document needs to be revised and re-approved, there should be an accepted way of identifying whether the extent of change to the document is major or minor. In the widely used "X.Y" versioning convention, X is reserved for major changes to a document and Y is reserved for minor changes to a document. The definition of a major change versus a minor change is subjective, and different organizations develop their own definitions of these terms (which is perfectly acceptable as long as there is clear understanding and agreement within the organization). Generally, a major change implies a significant change in the document to address known errors, omissions, or required significant clarifications. Minor changes pertain to typographical and grammatical errors, or required small clarifications.

A major change to a document can be reflected in the document version by incrementing the numeric digit *before* the decimal and resetting the numeric digit *after* the decimal to zero, e.g., version 1.4 -> version 2.0, and version 2.0 -> version 3.0. A major version number should not be skipped; e.g., version 1.4 -> version 3.0 is not permissible.

A minor change to a document can be reflected in the document version by incrementing the numeric digit *after* the decimal and leaving the digit *before* the decimal unchanged; e.g., version 1.0 -> version 1.1, and version 1.1 -> version 1.2.

Any real-life document change history consists of a sequence of major and minor changes as described above. A major change to a document can be followed by a minor change, and vice versa. Therefore, following would be a valid document change history: version 1.0 -> version 1.1 (minor change), version 1.1 -> version 2.0 (major change), and version 2.0 -> version 3.0 (major change).

The above example presents a simple scenario, wherein a document undergoes a first release, "version 1.0"; it is then rereleased with a minor change, "version 1.1." The document then undergoes a major change and it is rereleased as "version 2.0." However, real-life evolution of documents includes one additional small complexity. What typically happens is that once it is determined that "version 1.0" of a document needs to be revised, the document is revised and the revised document (which is considered a "draft") is reviewed in a meeting. The draft document is revised again (to address the comments received at the review meeting) before it is approved and released as "version 1.1." Therefore, there must be a way of identifying the "draft" document created *between* the approved versions 1.0 and 1.1. This can be done by appending "draft a" (or "draft b," or "draft c," and so on, depending on the revision level of the draft) to the document version that is planned to be approved (version 1.1). Let us study this with the aid of an example:

Consider the following document change history:

"Version 1.0" -> "version 1.1 draft a" -> "version 1.1" -> "version 1.2 draft a" -> "version 1.2 draft b" -> "version 1.2"

In the above sequence, the first approved release of the document, "version 1.0," was revised to "version 1.1 draft a." This version was reviewed at a meeting and reworked again before it was formally approved and released as "version 1.1." After the document was in use for some time, need arose to revise the document again. Therefore, version 1.1 of the document was revised to "version 1.2 draft a," which was reviewed at a meeting. It was agreed at the review meeting that the document needed to be reworked and rereviewed. The revised document, "version 1.2 draft b," was therefore rereviewed and reworked one last time before final approval and release as "version 1.2."

Instead of using the versioning convention described here, an organization may choose some other convention as per its needs. However, whatever versioning convention is used must facilitate quick identification of the revision status of the document. It is preferable that the document version also provide a quick means of identifying a draft document versus an approved document (unless some other means is used to communicate a document's approval status). For example, "Version 1.1 Draft" clearly indicates the document is a draft document (not yet approved for use), while omission of the word *Draft* from the document's version number indicates that the document is approved for use. With a document revision, it is preferable that the document version also indicate whether the document has undergone a major revision or a minor revision.

3.2.5 Document Content

Following are some guidelines regarding document content (for examples, refer to sample QMS documents in the Appendices):

1. All QMS documents and records should have the same "look and feel." Consistency in templates and forms can be ensured by following some general guidelines, such as:
   - The same header and footer on all templates and forms, with information such as:
     - Document title
     - Document number
     - Document version
     - A statement to indicate the company proprietary nature of the document
   - Each QMS document, when appropriate, should use the same title page, containing the following:

   - Standard information, (e.g., organization's name and logo). The title page also can include a unique logo to indicate that it is a QMS document. This enables quick identification of QMS documents.
   - Customizable information, (e.g., document title)
2. Each QMS document or record should contain content as per the guidance contained in the associated form or template. Each QMS document must identify its purpose and scope clearly. Correctness of the document content should be reviewed and enforced during document review meetings.
3. Procedure and work instruction length should comply with recommendations made earlier in this chapter.
4. Deletions from the template and forms should not be allowed. If a particular section is not required, it should be marked "not applicable."
5. Insertion of additional sections in a document created by using a standard form or template should be allowed when necessary; however, certain rules should be established to handle addition of new sections. An acceptable rule may be that new sections can be added only after the last section in the template or form. Similarly, a new subsection might be added only after the last subsection (in the relevant section) in the template or form. For example, if a document template has five sections (see Figure 13), and a new section needs to be added in a document created by using this template, it can only be added as section 6. Similarly, if a new subsection needs to be added under section 4, it can only be added as subsection 4.3. This approach helps maintain consistency across documents of the same type, which would be compromised if authors were allowed to deviate from the sequence of sections included in the standard template or form.

3.2.6 References and Related Documents

QMS documents, when appropriate, must contain a section listing all the documents used as a reference while creating the document or related to the content of the document. This helps point the user to relevant recommended readings associated with the document.

3.2.7 Document Review and Approval

All QMS documents, with the exception of records, must be reviewed and approved before use. A document review must involve *all* stakeholders, but the required approvals for the document must be limited to only the

Document XYZ

Section 1: Purpose

Section 2: Scope

Section 3: Terminology

Section 4: Description

Section 4.1: Activities

Section 4.2: Activity Measurements

Section 4.3: Inputs & Outputs ← New Section

Section 5: Related Documents

Section 6: Training Requirements ← New Section

Figure 13 Handling addition of new sections in a document created from a standard template

key stakeholders. Caution should be exercised in determining the number of required approvals for a document, because failure to limit these to a small number can result in an unnecessarily long list of required approvals that renders the entire document approval process inefficient and introduces unnecessary delays. A good general rule of thumb to follow for identifying required approvers for a QMS document is to include the following as approvers:

- The department that authored the document;
- The prime department which provides information that serves as input for the document (primary producer); and
- The prime department that uses the document (primary consumer).

3.2.8 Document Change Requests

The department that owns a QMS document should be the only one authorized to make changes to the document. However, a mechanism must be available to all employees to request a change to an approved QMS document. A document change request form should be available for submission of document change requests. (A sample document change request form is provided in Appendix D.3.) The submitter of a document change request should indicate for which document a change is requested, and the submitter should provide a brief summary of the proposed changes along with a reason for the requested changes. The PMC and the quality assurance department should review all submitted document change requests periodically. Minimally, the respective process owner (or document author) and the quality assurance department should review the submitted document change request. The document author should implement approved change requests within a reasonable time frame. In case a document change request is rejected, the request should be formally closed only upon discussion and agreement with the request submitter. An escalation mechanism might be required for handling of disagreement between the request submitter and the reviewers of the document change request. For example, disagreements may be escalated and resolved between the request submitter's supervisor and the management representative.

3.2.9 Document Change Review and Approval

Changes to an approved QMS document must be reviewed by personnel in the same roles as the original reviewers of the document. Because employees in an organization may be transferred or promoted, or may leave the organization, it is not advisable to require that changes to approved documents be reviewed by the *same personnel* who reviewed the previously released version of the document. Similarly, when appropriate, the revised document must be approved by personnel in the same roles as the approvers of the original version of the document.

3.2.10 Emergency Changes to Previously Released Documents

A process should be established for handling emergency changes to a previously approved QMS document. This process occasionally might be required to correct critical inaccuracies and/or deficiencies in an approved QMS document immediately. The emergency change process is intended to be an expeditious yet temporary way of performing a document change, as compared with an organization's standard process for document change, which may be more time-consuming. Such an emergency change process is especially important in the manufacturing industry, where immediate corrective or preventive action might need to be implemented in a process. An emergency change process may be as follows:

The submitter of the document change discusses the proposed change with the document's author. If the two parties agree, a hard copy of the document is redlined (to indicate the change), initialed, and released by the author for immediate use by the impacted personnel. Immediately thereafter, the author initiates the organization's standard process for implementing the required document change. Following this process, once the document has been formally revised, reviewed, approved, and released as the next version, the original redlined copies that were circulated for temporary use are withdrawn from circulation and considered obsolete.

3.2.11 Document Change Identification

All documents should contain a "change history" table that briefly summarizes the changes included in each formally approved and released version of the document. Note that the change history table need not list draft versions of the document. A document may evolve through a number of draft versions; including each draft version in the change history table may make the table unnecessarily long without adding much value. Moreover, the users of a document usually are interested in knowing what changes were included in each approved release of the document. An example of a change history table is shown in Table 12.

It is also desirable that the revised document use change lines (or bars) to indicate which paragraphs of the document have changed. A change line is a vertical line in the margin of the document, adjacent to the text that has been changed. Change lines help in quick identification of sections of the document that have changed. This feature is available in all widely used word processing software. In addition to use of change lines, text insertions in the latest version of a document may be highlighted in a distinct font, and text deletions may be indicated with text strikethroughs.

3.2.12 Notification of Document Change or Document Release

In the event of changes to a previously approved document, or in the case of a new QMS document, appropriate departments should be notified once the document is approved and released. For this purpose, a periodic notification, e.g., on a weekly or biweekly basis, is acceptable (unless the change is of an emergency nature, as described earlier).

Generally, information regarding departments that might be affected by a QMS document (and therefore need to be notified of its release) can be obtained from the document owner. In the case of procedures, this information can be obtained directly from the procedure by referring to the *applicability* section. (Refer to Appendix D.1.) Such a notification may be sent electronically (such as by electronic mail), or by an interdepartmental memo. In the case of changed documents, the notification should contain a brief summary of the change. In the case of newly released documents, the notification should describe the purpose, scope, and applicability of the document.

Table 12 Document Change History

Document Version	*Date*	*Change Performed By*	*Summary of Change*
Version 1.0	8/1/2002	Bob Ingram	First approved release of document
Version 2.0	10/15/2002	Deb Spitz	Added new section 4.7 to describe product safety requirements
Version 2.1	11/29/2002	Deb Spitz	Reformatted figure 4 and added subsection 4.3

3.2.13 Master List of Controlled Documents

The organization should maintain a master list of QMS documents that identifies the current approved versions of controlled documents. In case of electronic implementation of the QMS documentation repository, the hypertext list of current approved versions of QMS documents available in the repository can serve as the master list of controlled documents. In case of hard copy implementation of the QMS documentation repository, a master list similar to the one shown in Table 13 may be used. Note that in that case, the Document Location column in Table 13 would refer to the location of the master copy of the controlled document. (As stated previously, this location can be centralized or distributed.)

Just as in the case of QMS documents, because the number of records in an organization typically is large, the quality assurance department can maintain a master list of QMS-related records to serve as a quick reference. An example of such a records master list is shown in Table 13. Such a master list serves as a handy reference for control of records, as well as for use in internal quality audits. Alternatively, information regarding record location, retention time, and disposition may be recorded in the appropriate procedures and work instructions.

3.2.14 Handling of Obsolete Documents

Once a document becomes obsolete, it must be withdrawn from use immediately. If the document was distributed electronically, say by means of an intranet Web site, the obsolete document can be withdrawn by removing it from the Web site. If the document had been distributed by means of hard copy circulation, each distributed copy of the document must be withdrawn. In either case, if the obsolete document is superseded by a new version, the new version of the document must be properly distributed.

Table 13 Records Master List (partially complete)

Record Name	*Document Location*	*Retention Time*	*Disposition State (Dispose/Keep)*
Contract review records	Legal department filing cabinet	5 years	Keep
Product Test reports and results	QMS repository	5 years	Keep
Project document review records	QMS repository	2 years	Dispose

3.2.15 Authorization for External Release of Internal Documents

Rules governing the external release of QMS documents must be put in place. This includes release of QMS documents upon special request by key customers. Acceptable release authorization may include management-level approval from the department that owns the document to be released. Before releasing the document, the owning department should consider purging the document of information that may be deemed confidential or otherwise unsuitable for release to an outside party.

3.2.16 Control of Documents of External Origin

The organization should ensure that all documents of external origin used within the scope of the QMS are clearly identified, and their distribution controlled. It is desirable to have a master index of all documents of external origin in use, along with a reference to each document's location. Such documents may be assigned a document number with a unique identifier to clearly indicate their external origin. A sample master list of documents of external origin that are numbered in accordance with the document numbering convention in Table 10 is shown in Table 14.

3.2.17 Document Storage and Publication

The final element of documentation management and control pertains to storage of master copies and publication (or dissemination) of QMS documentation. Mechanisms must be in place to ensure that approved master copies of QMS documents are stored in a secure location, and that QMS documents are made available to relevant employees. This may be accomplished by establishing a:

Table 14 Master Index of Documents of External Origin

Document Number	*Document Name*	*Document Version*	*Date*	*Document Location*
EXT-ST-9001-0234	ISO 9001:2000 official copy	Year 2000	12/15/2000	Quality assurance department
EXT-GD-INPR-0346	Internet protocol implementation guideline document	6.0	5/14/2001	Engineering department

- *Hard-copy QMS documentation repository,* wherein the approved master hard copy of each QMS document is stored in a secure filing area (indexed by document number), and copies are distributed to departments in hard copy process binders; or
- *Electronic QMS documentation repository,* wherein the approved master copy of each QMS document is stored in a secure database. The documents are made available electronically to personnel, such as by publishing the QMS documents on an intranet Web site.

Alternatively, a combination of the two approaches may be used wherein approved hard copies are stored in a secure filing area and the matching soft copy of the document is made available electronically to personnel.

Electronic storage and distribution of QMS documents is becoming increasingly popular due to its sheer ease of use and maintenance. Hard-copy QMS implementation, on the other hand, generally entails:

- Higher personnel costs due to manual handling of QMS documentation. This includes resources required for filing of master copies, and creating copies for distribution to personnel. Similarly, when a QMS document is revised or withdrawn from use, obsolete copies must be physically withdrawn from use.
- More time required for circulation of QMS documentation due to manual storage and distribution
- More errors and inaccuracies, because these are inherent in any manual process of storing, copying, and distributing documents

Insofar as the organization of information in the QMS documentation repository is concerned, Figure 14 shows an example of how an electronic

QMS documentation repository (assumed to be implemented in a corporate intranet) may be organized. Each bullet item is intended to be a hypertext link to the relevant information. Note that the structure of the QMS documentation repository shown is consistent with the logical structure of the QMS documentation hierarchy in Figure 6. If an organization uses a hard-copy QMS documentation system, the process binders typically contain Level 1, Level 2, and Level 3 documentation (including relevant departmental documents only), and any additional useful information as appropriate, such as a form to submit improvement suggestions or a QMS glossary. Project documentation and records (Level 4 documentation) is maintained in separate project binders and databases.

3.3 QMS DOCUMENTATION PROCESS

Now that all the necessary elements of a documentation management system have been described, you are ready to begin creating QMS documents. What documentation process should be followed to create, review, and approve QMS documents? This section describes a *high-level* documentation process that ties together some of the key elements of document management and control explained earlier in this chapter. A specific application of such a documentation process might be for creating QMS procedures (refer to Appendix C.1), and for creating project documents.

Step 1: Identify suitable document author — The first step is to identify a document author who possesses appropriate subject matter expertise. Typically, a management person from the function tasked to create the document selects a suitable document author.

Step 2: Create draft version of the document — The document author creates a draft version of the document by using the applicable template (if one is available).

Step 3: Review the draft document — Once a draft version of the document has been prepared, it is circulated for review to appropriate reviewers (or functions) that are considered stakeholders in a document. The review may be in the form of an informal and/or formal review. (Refer to Formal and Informal Reviews Procedure in Appendix C.4.) The reviewers are provided a reasonable amount of time to review the material. Depending on the type of review, the reviewers might provide their feedback either in private to the author or at a review meeting.

Step 4: Rework the document as per reviewer feedback — After receiving the feedback from the reviewers, the author reworks the document in accordance with comments provided by the document reviewers. Note that this cycle of review and rework of the document might undergo several iterations until the reviewers are satisfied that the document is

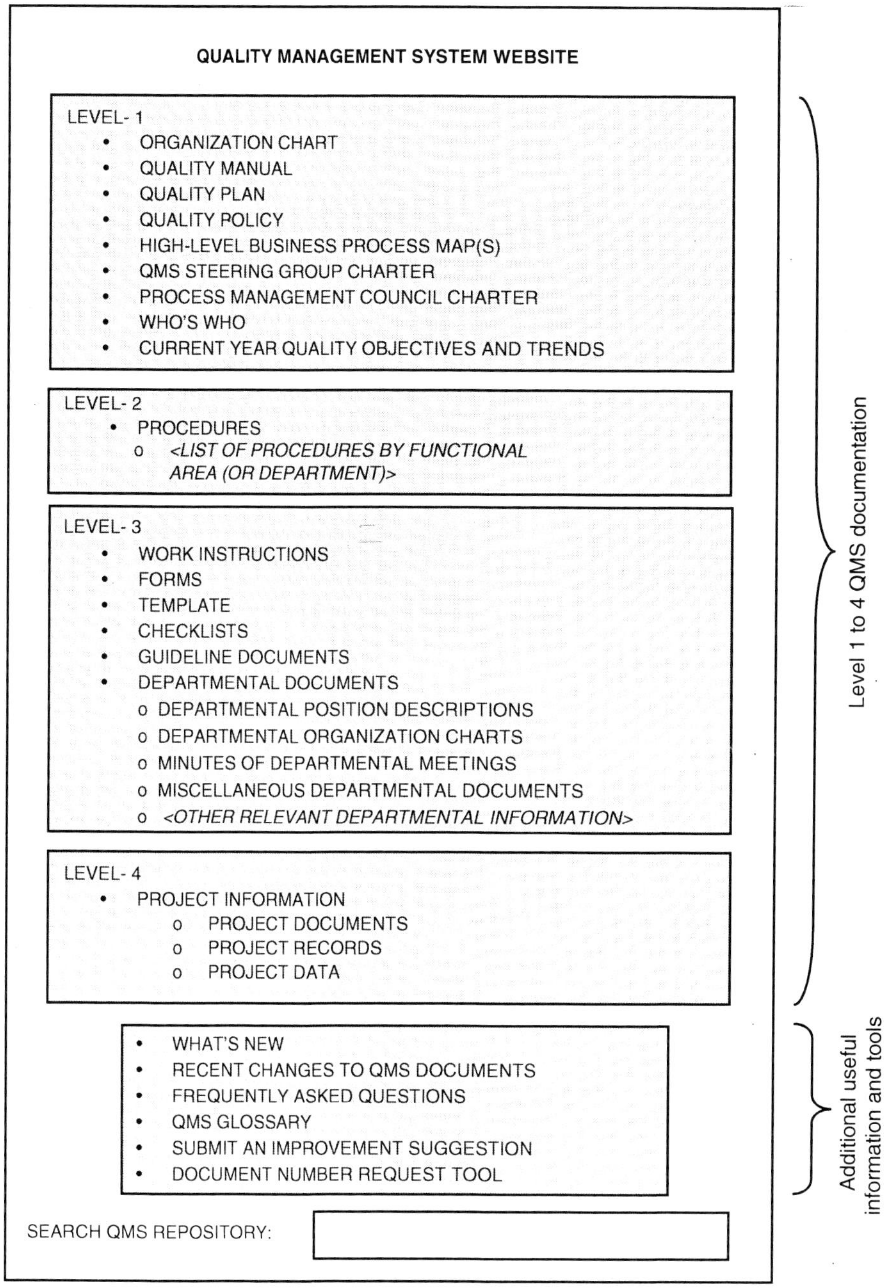

Figure 14 Conceptual Layout of a QMS Documentation Repository Intranet Home Page

ready for approval. Therefore, the document may evolve through several draft versions before it is considered ready for approval.

Step 5: Approve and publish the document — Once the document rework is complete, and it has been determined that a rereview is not required, the document is circulated for approval to the identified approvers. Once all the required approvals have been obtained, the author submits the master copy of the document along with the review record to the document controller. The document controller stores the document in the controlled repository and makes copies available for use.

PHASE II

DEFINE

4

DEFINING ORGANIZATIONAL PROCESSES

With the start of the QMS definition phase, the process of defining and documenting the QMS begins. For companies implementing a QMS to comply with a specific QMS standard, quality requirements are primarily *externally* specified (and complemented by *internally* specified quality requirements). Members of an external agency formulate quality requirements that are published in an approved national or international QMS standard. These external quality requirements are complemented by internal quality requirements that are specified by the respective company as per its unique needs and expectations. In the case of companies that are not pursuing compliance to a QMS standard, all quality requirements are internally specified. However, such companies usually refer to relevant QMS standards, industry-specific best practices, and generally accepted quality best practices (such as Malcolm Baldrige National Quality Award criteria) when formulating internal quality requirements. Generally, most internal quality requirements are specified as *shall* and *should* statements in the QMS documentation.

In both cases, the process of defining the QMS is fundamentally the same, and comprises the following activities:

First, determine the current situation in the organization by performing an initial assessment or gap analysis (against a QMS standard, if one is in use). Use this information to fine-tune the implementation plan and improve its feasibility by better accommodating the current ground realities. Then, prioritize the quality deficiencies uncovered during this exercise and immediately address those that are deemed critical and require immediate corrective action. Additionally, deficiencies regarded as low-hanging fruit may also be addressed at this time; these are deficiencies that can be fixed relatively easily (at least temporarily) by minor process and operational changes, and that translate in immediate and significant benefits. Such benefits may include but are not limited to an improved product (or service quality), improved operational efficiency and/or effectiveness, and improved customer satisfaction.

Second, map and define the organizational processes in a top-down fashion, and create supporting process assets. In defining the processes, specify appropriate internal quality requirements, and ensure that defined processes are capable of satisfying all applicable requirements (internal and external).

Each of these activities is examined in detail in this chapter. Guidance on fundamental quality practices that are relevant to various functional areas in product development companies is provided in Chapter 5.

4.1 IDENTIFY AND CLOSE CRITICAL QUALITY GAPS

4.1.1 Analyze Quality Management System Standard Requirements (if applicable)

The first step in determining the current situation is defining the desired state as precisely as is possible at this stage. The desired state represents the benchmark (or reference) against which the current state of the organization will be assessed.

For a company that is pursuing implementation of a QMS standard, the task of identifying the desired state is relatively trivial, because the requirements in the applicable QMS standard provide an accurate definition of the desired state.

On the other hand, for a company that is not pursuing implementation of any QMS standard, the definition of the desired state is somewhat less precise. Since the company has yet to embark upon the definition of its QMS, internal quality requirements are not yet specified. However, it is possible for such companies to perform an initial high-level assessment of the current state against *relevant* QMS standards' requirements, quality best practices, and industry-specific best practices. This is because many of these requirements and best practices, having been deemed relevant to the company, eventually would be formulated into internal quality requirements. For the purpose of conducting a high-level assessment, a set of QMS standards requirements and best practices relevant to each functional area of the organization should be identified (along with the source) in a gap assessment matrix similar to the QMS traceability matrix shown in Table 15. This is meant to be a preliminary set, and is subject to change once QMS definition gets under way. This set of quality requirements and best practices therefore provides an approximate representation of the desired state against which the company can be assessed.

4.1.2 Perform Preliminary Gap Analysis

The second step is to perform a preliminary assessment (or gap analysis) of the current state of the company against the desired state identified in the preceding step. First, rate the adequacy of current practices as fully compliant, partially compliant, or noncompliant against the desired state. The outcome of this exercise defines the current state. Next, for each requirement (or best practice), identify the difference between the current state and the desired state. This constitutes the quality gap that the QMS implementation seeks to address.

4.1.3 Revise Implementation Plan

Once the quality gaps have been identified, analyze the list of gaps for severity, urgency, and anticipated effort, and prioritize their implementation accordingly, assigning the implementation to appropriate process owners (and/or management personnel). Use the information gained from the preliminary gap analysis to revisit and revise (as appropriate) the basis of estimates, assumptions, anticipated task durations, planned activities, and related implementation planning information. For example, if the preliminary assessment discovers that some of the required QMS documentation already exists, then it is generally preferable to reuse existing documentation (and enhance it, as needed) instead of creating new QMS documentation from scratch. Revise the QMS implementation plan accordingly to reflect reuse of existing QMS documentation, where appropriate. Reuse of existing QMS documentation offers the benefit that personnel are already familiar with it (assuming the documentation is in use). They are more likely to accept incremental changes to existing QMS documentation than radical change (new QMS documentation written from scratch) — although in some instances, the latter may be preferable. Reuse of existing QMS documentation also offers time and cost savings.

4.1.4 Correct Critical Quality Discrepancies

Most of the gaps discovered in the preceding step should be addressed during the course of definition and deployment of the relevant processes. However, there inevitably will be certain quality gaps that will be deemed to be of critical severity and high urgency, or that are relatively easy to fix and offer significant benefits. The appropriate process owners (and/or management personnel) should address such quality gaps immediately in consultation with appropriate personnel from other organizational functions who might be affected by the changes. Some of the solutions adopted may be temporary and subject to change once the relevant processes are examined and defined later in the QMS definition phase.

4.2 ESTABLISH AN ORGANIZATIONAL PROCESS BASELINE

Once the current situation in an organization has been determined and performance gaps requiring immediate closure have been addressed, the task of mapping business processes and creating supporting process documentation can begin. For reasons explained earlier, process mapping and process definition should proceed in a top-down fashion, beginning with a high-level organizational process map (or a set of high-level organizational process maps), then drilling down deeper into each process. The set of the organization's processes and their documentation in the form of process maps and procedures collectively serve as the organizational process baseline.

An organizational process baseline helps forge a common understanding of how tasks are performed (or are supposed to be performed), and it guides the execution of all processes and activities in the organization. It helps managers plan the execution of their activities (in accordance with the defined processes), and it enables comparisons of actual process execution against the defined processes to identify deviations and inconsistencies. The capability and performance of the organizational process baseline should be monitored continuously (and measured, when appropriate) to identify opportunities for process improvement. Refer to Chapter 8 for discussion on different mechanisms that may be used for continuous process improvement.

4.2.1 Define High-Level Organizational Processes

When defining organizational processes, some questions arise and must be answered:

Identification: What processes need to be defined? What are the organization's high-level business processes?
Scope: What is the scope (or boundary) of each process to be defined?
Participants: Who should participate in process definition?

Approach: How should each process be defined (i.e., what steps should be followed for defining the processes)?

Level of detail: To what extent should the identified processes be defined?

These questions are examined next.

The task of identifying and defining an organization's business processes should be performed incrementally in the following steps:

Step 1: Identify all the functional areas in the organization (typically delineated in an organization by "departments").

For this purpose, the organization's business operations are decomposed at a high level into major functional areas, such as product marketing, product development, and customer support. This list of functional areas in an organization can be obtained from the organization chart (if one is available). At this point, because the organization does not yet have a high-level process map of its business operations, each of these functional areas is, in essence, a black box. For illustration purposes, Figure 15 shows what such a high-level functional decomposition of a product development company's business operations may look like.

Step 2: Examine how the identified departments collectively execute the organization's business operations by mapping the interplay of key business processes.

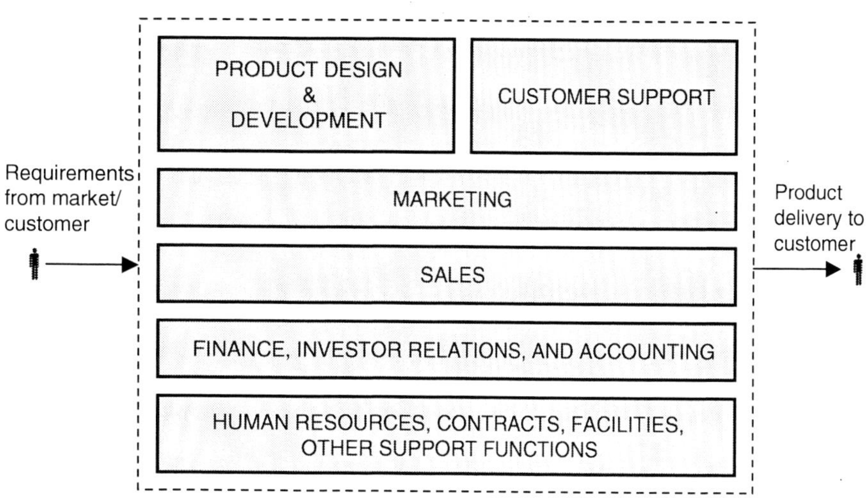

Figure 15 Major functional areas of a product development company

The task of identifying and mapping an organization's business processes should encompass intradepartmental and interdepartmental processes. At the highest level, this entails generating a process map (or set of process maps) encompassing key technical, management, and support processes that translate market (or customer) requirements into product requirements, followed by design, development, and delivery of a finished product to the customers.

In order to generate such a process map, a process mapping workshop should be organized with appropriate cross-functional representation from relevant departments (hereafter referred to as the process mapping team). Generally, at this stage, participation of management personnel in the process mapping team is sufficient, and participation of practitioners should be increased gradually as the organization begins to decompose the metaprocesses into subprocesses.

The process mapping workshop should be coordinated and facilitated by the management representative, with the assistance of the corporate quality assurance department personnel as needed. It is highly recommended that this process mapping exercise be performed at an offsite location, so that the process mapping team can work uninterrupted. The room used for this activity should have a large white board to map the process by affixing uniquely colored Post-it™ notes representing processes, activities, and inputs/outputs to the process map. The process mapping team can adopt one of the following two approaches for process mapping, or it may choose to use a combination of the two approaches.

Approach 1 — Representatives from each department in the process mapping team are provided with different* colored Post-it™ Notes (hereafter referred to as sticky notes). The process mapping team is first instructed to consider its end-to-end metaprocess as a black box (see Figure 16) — the scope of this process spans from the receipt of market or customer requirements (or any other external input) to delivery of the finished product. The team is then asked to specify, as completely as possible, the different types of external inputs that may be received at the beginning of this metaprocess.

* Different departments are provided with different-colored sticky notes so that they can be visually distinguished in the process map. Additionally, if the number of different colors of sticky notes is fewer than the number of departments, a unique annotation may be used on the sticky notes (to identify the associated department), so that the same color may be used for more than one department.

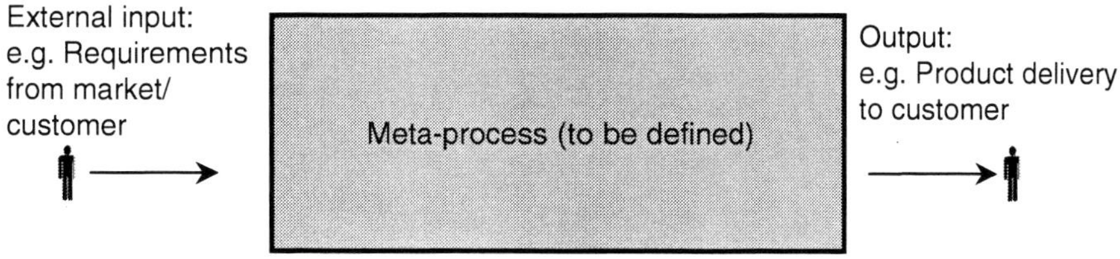

Figure 16 Initial black box representation of the metaprocess (prior to process mapping exercise)

Examples of different types of external inputs include but are not limited to:

- Request for information (RFI)
- Request for quotation (RFQ)
- New product requirement
- Request for a new product
- Report of product defect

Each of these inputs does not necessarily result in the same output. For example, the output in response to an RFQ is generally a price quote for the requested product. However, a new product requirement generally results in the requested product capability being delivered (after successful technical feasibility review, and contractual and price negotiations). Depending on what external input is received, and what output is required, different departments in the organization get involved in processing the received inputs. For example, an RFQ typically requires the involvement of personnel from only the sales department. However, handling of a new product requirement requires wider cross-functional participation — initially, for pricing and contractual negotiation, and then for the design, development, delivery, and support of the new product capability.

For each type of external input, the process mapping team should peer inside the metaprocess and trace the processing of the input through all the intervening processes, until the final output is produced. The team begins the process mapping exercise by writing the name of the external input on a uniquely* colored sticky note and affixing it to the wall. Next, the department that first processes the external input on receipt writes

* A separate color should be reserved to represent all process inputs and outputs so that they can be visually distinguished from processes (or subprocesses) in the process map.

the name of the process it executes* on its colored sticky note, which is affixed after the sticky note representing the external input. Outputs produced by this process (intermediate output) should also be represented with separate sticky notes for each output, followed by the next process, tracing through the entire process until the final output is mapped. During process mapping, the team generally should adhere to the following guidelines:

- All processes and intermediate outputs should be represented in chronological order.
- All intermediate outputs should be represented in the process map.
- In order to reflect the flow of information, each process should be connected to its inputs and outputs using arrows.
- Avoid crossing of lines (arrows) in the process map.
- Ensure that the process flow is clear and uncluttered.

Because this process map is intended to be only a high-level representation of the organization's metaprocess, details of individual processes need not be elaborated at this stage. Instead, logically related activities should be grouped together as a process and, when possible, represented on the process map by a single sticky note. The process map at this stage therefore is the same as a macro-level flowchart of the organization's metaprocess, wherein all key processes are depicted, but not decision points. Later, as the individual processes are defined, there is a need to represent all activities and decision points using traditional flowcharts. (Refer to flowchart symbols in Appendix D.1.) As one steps down the process hierarchy of an organization, the information contained in the corresponding process flows — process maps at the higher level, and flowcharts at a lower level — accordingly increases.

Figure 17 shows what the results of a process mapping exercise for a metaprocess might look like. In the figure, all inputs and outputs are represented by sticky notes of the same color, and different colored sticky notes for the processes indicate that those processes are owned by different departments.

Once a high-level process map has been constructed, the process mapping team should document pertinent high-level process information in the sticky note for each process. Such information includes a precise

* Note that a subprocess may be intradepartmental or interdepartmental. Therefore, the color code of the sticky note represents the department that has overall responsibility for the subprocess, and it is *not* intended to imply that the subprocess is wholly executed within that department. Refer to the explanation of process ownership in Chapter 2.

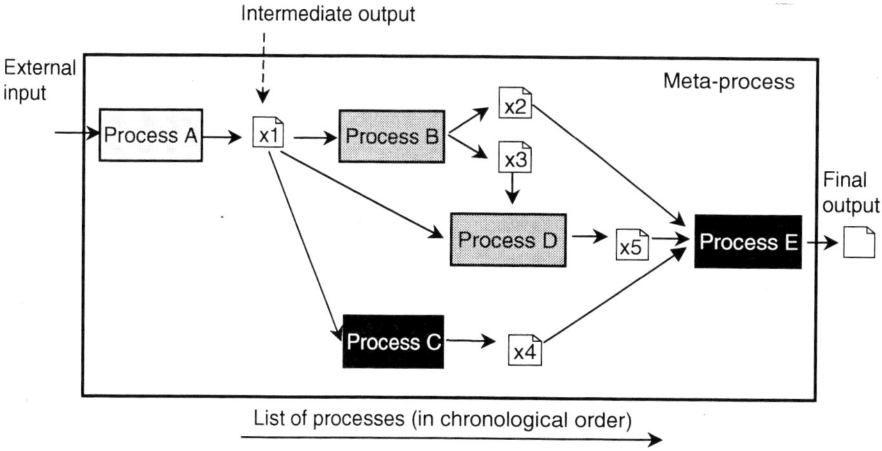

Figure 17 Results of Process Mapping Exercise for the Metaprocess

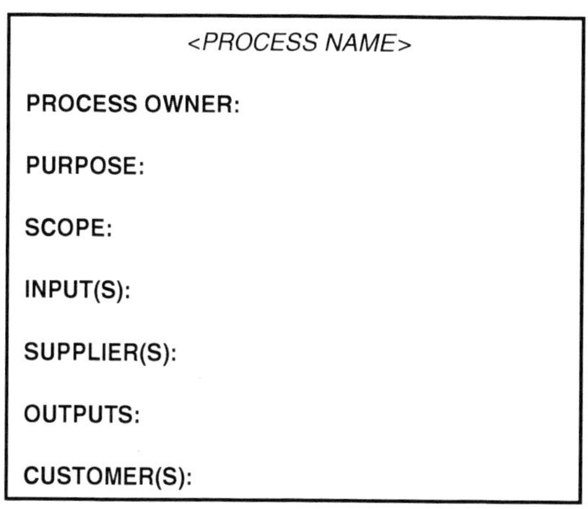
<PROCESS NAME>

PROCESS OWNER:

PURPOSE:

SCOPE:

INPUT(S):

SUPPLIER(S):

OUTPUTS:

CUSTOMER(S):

Figure 18 Sticky Note Information Requirements

definition of the purpose and scope of each process, list of inputs and outputs, input providers (suppliers), output receivers (customers), and identification (by name) of the process owner (see Figure 18). Figure 19 shows a sticky note completed with pertinent information for process B in Figure 17.

Once the sticky notes have been completed to include the process information, the facilitator of the process mapping workshop should transcribe the complete process map and information from the sticky notes to a formal document that captures the results of the workshop. Copies

SUBPROCESS B

PROCESS OWNER: Mr. XYZ

PURPOSE: The purpose of this process is to ...

SCOPE: This process includes ...

INPUT(S): X1

SUPPLIER(S): Process A

OUTPUTS: X2, X3

CUSTOMER(S): Process D (X3), Process E (X2)

Figure 19 Sample Information for Process B

of this document should be distributed to all process owners so that they may use the preliminary process information for their respective processes as a starting point for subsequent flowcharting and elaboration in onsite meetings. (Refer to the next section, "Define low level organizational processes.")

Approach 2 — The second approach to process mapping is slightly different in that the process map not only consists of processes that are chronologically ordered, but also are ordered by the functional area (department) that owns each process (refer to Figure 20). The process mapping team begins by listing all the departments that are involved in the metaprocess along the left-hand axis on the white board, and creating a horizontal band for each (by drawing dotted lines). Process mapping then proceeds as in the first approach, but with each process located within the horizontal band of the department that owns the process. For example, in Figure 20, the location of processes B and D in the horizontal band for FUNC-2 means that the FUNC-2 department owns these two processes. Such a process map provides the same information as the one in the first approach, with the overall metaprocess better organized. Because the purpose of the horizontal bands is to show how the different departments are involved in the metaprocess, there is no need for uniquely colored sticky notes to denote different departments (although they may still be used).

Alternatively, the process mapping team may adopt a combination of the two approaches, wherein the process is first mapped using the first approach, and then the process map is reorganized according to the second approach.

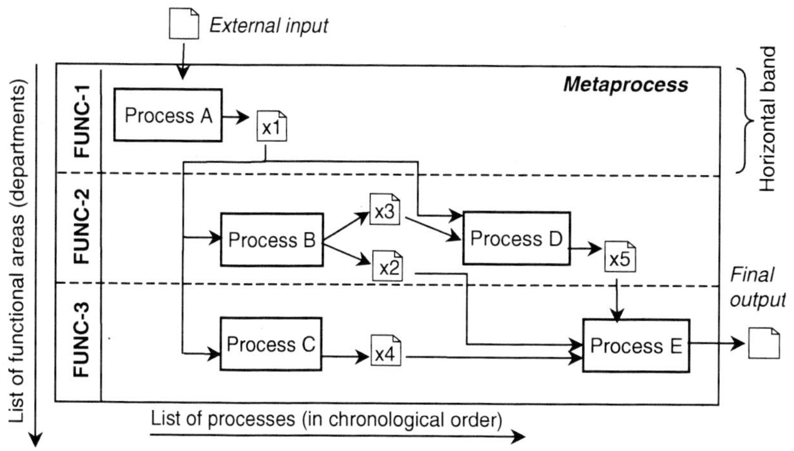

Figure 20 High-Level Business Process Map Showing the Role Played By Different Departments (Func-1... Func-3)

Let us consider another example of a process map (see Figure 21). The figure illustrates the product development process at the highest level of abstraction (least detail) for a software development company. The company's product development process begins with the requirement definition process, which entails collecting product requirements from the customers and other inputs. The collected requirements are documented in a "product requirements document," and the document is reviewed within the company and with customers, as needed (see Figure 22). The defined requirements are then analyzed to create one or more specifications that specify detailed software requirements. The reviewed and approved specifications serve as input for the design process, which entails specifying software design in compliance with the software requirements specifications. Once the software design is complete, the software code is written, verified, and integrated by the software engineers prior to handoff for software validation. After the software validation results meet predefined quality criteria, the software is considered acceptable for release and the product delivery process is initiated. Note that the process maps in this example are divided into two levels of detail: an initial high-level process map in Figure 21, followed by an example of a detailed process map in Figure 22. This is to illustrate that process maps need not be limited to high-level processes; they also can be used for lower-level processes. Alternatively, an organization may choose to include the level of detail provided in Figure 22 within the process map depicted in Figure 21, then create flowcharts for each of the subprocesses as explained previously. The latter approach is acceptable if it does not result in an overly complex process map that is difficult to understand.

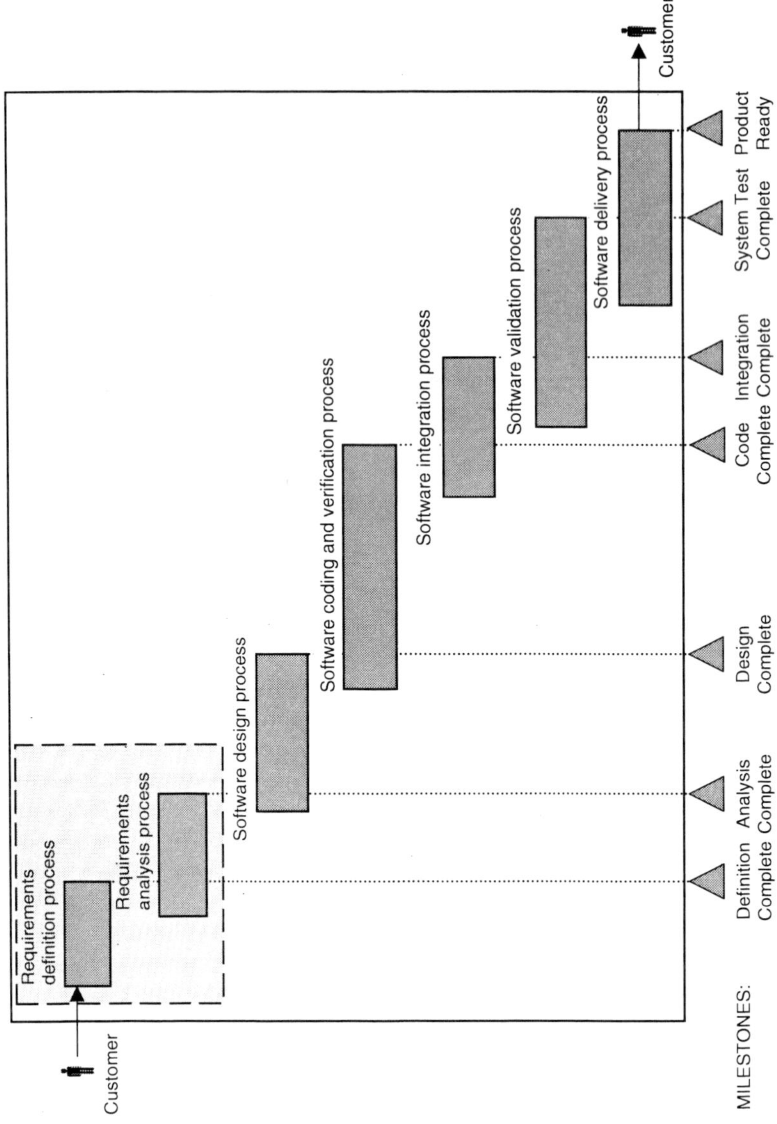

Figure 21 Product Development Process Map for a Software Development Company (First level of detail)

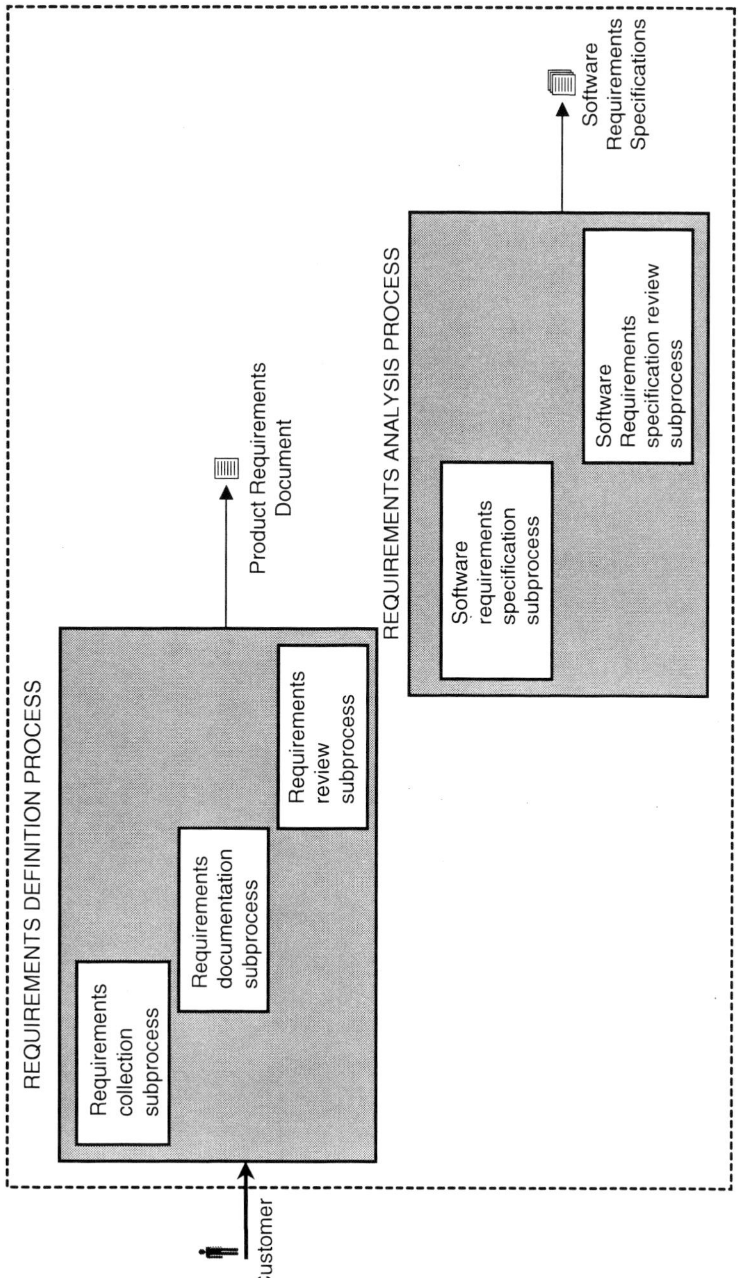

Figure 22 Requirements Definition and Analysis Processes (second level of detail)

4.2.1.1 *The Case for Process Improvement during Process Mapping*

The foregoing explanation of process mapping described how an organization can map its current business processes. It is quite natural that during the process mapping exercise, the process mapping team will uncover and discuss deficiencies in the process in the form of redundant steps, unnecessary time delay, irrelevant outputs, or other problems encountered during process execution. Therefore, the process mapping team should take advantage of this opportunity to make *incremental* improvements to the process. Once the team has mapped (and recorded) the process as it is currently executed, it can make slight modifications to the process map to address known deficiencies. The modified process map then describes how the process should be executed. Such slight modifications should translate to only minor operational changes that are acceptable to the process mapping team. However, if the existing process has serious deficiencies that require redesign (process redesign), or if the existing process is so seriously and fundamentally flawed that it requires a fundamental rethinking and redesign from scratch (business process reengineering), such process improvements should be discussed *after* the initial process mapping exercise. Otherwise, it would distract the team from the very purpose of the process mapping exercise — to examine the organization's business operations by understanding the interplay of its current business processes.

4.2.2 Define Lower-Level Organizational Processes

Once the metaprocess of the organization has been defined in the offsite process mapping workshop, the task of defining the lower-level processes can begin. At this stage, definition of the lower-level processes generally takes the form of process flowcharts that depict all process activities and decision points, unless the organization has several levels of process hierarchy, in which case process flowcharts are usually appropriate at the lowest level and process maps should be used to represent higher-level processes. The task of coordinating the definition of the lower-level processes should be assigned to the respective process owners (who were identified during the high-level process mapping exercise), and lower-level processes may be defined in meetings conducted onsite. The following parties should participate in the definition of lower-level processes:

- Management and practitioner-level representatives from the departments that execute the process
- Representatives from departments that provide inputs to the process
- Representatives from departments that receive output from the process

Such cross-functional participation in process definition enables agreement on issues such as:

- Interface objects (inputs and outputs)
- Content of those interface objects
- Expectations regarding verification to be performed by the producer of the interface object prior to its release
- Expectations regarding when the interface object should be made available
- Who needs to approve the interface object prior to its release

As an example, let us assume that process B in Figure 20 is an intradepartmental process, completely executed within one department, FUNC-2. Representatives from FUNC-2, representatives from the department that produces input 'x1' (i.e., FUNC-1), and representatives from the departments that receive outputs 'x2' and 'x3' (i.e., FUNC-2 and FUNC-3) should participate in defining process B. The lower-level processes may be flowcharted using the same sticky note process mapping method described in the previous section. The only difference is that instead of mapping processes within a metaprocess, activities within each process (along with decision points) are mapped.

4.3 CREATE ADDITIONAL PROCESS DOCUMENTATION

Although process maps are valuable process documentation for an organization, they generally do not provide adequate guidance to process practitioners for consistent process execution. Therefore, process maps usually need to be supplemented with additional documentation to provide pertinent process information. Such additional information includes specification of process entry and exit criteria, description of tools, methods, techniques, competency requirements, roles and responsibilities, and standardized format and content for process outputs (templates and forms). This additional process information should also be documented in the QMS definition phase.

The process owner has overall responsibility for identifying required process documentation and creating the identified process documentation by the due dates (as explained in Chapter 2; due dates for QMS documentation are recorded in Table 6). Generally, the task of writing process documentation should be distributed among several qualified department personnel (from one or more departments involved in a process), as opposed to burdening one person with authoring all process documentation. Broad employee participation in the creation of process documentation and in subsequent cross-functional reviews of the documentation prior to approval also fosters a sense of active involvement and QMS ownership, which in turn helps secure employee buy-in.

The process owner is responsible for coordinating with management personnel from the relevant departments for the selection of qualified personnel for creating the process documentation. The following is a list of desired skills that the process owner and the respective department management personnel can use as criteria for selection of personnel for this purpose:

- Well-rounded education and business process background
- Sufficient experience with the process, either in a management role, or as a process practitioner
- Sufficient cross-functional understanding and familiarity with interfacing processes
- Diverse experience in the specialty, in organizations of different sizes and maturity
- Strong communication skills (verbal as well as written)
- Strong commitment to quality and continuous process improvement

There are three steps in creating new process documentation:

1. Specification of document format
2. Definition of document content
3. Review and approval of document format and content by appropriate stakeholders

All stakeholders in a process — those executing the process, those receiving outputs from the process, and those seeking assurance and controls for the faithful execution of the process* — should be satisfied that their specific needs have been addressed in the relevant process documentation. The needs of the various parties constitute the set of requirements that the documentation must meet prior to approval. Some of the requirements that the documentation must meet are stipulated formally, while others can be expressed informally. For example, the needs of internal stakeholders usually are expressed informally in meetings, reviews, or other informal interaction. On the other hand, requirements of external parties such as customers usually are specified in contractual agreements, requirements documents, or formal minutes of customer–supplier meetings; requirements of external auditors, such as registrars, are derived from the applicable industry QMS standards document; and

* This includes internal parties, such as quality assurance personnel and management personnel, or external parties, such as customers and external auditors.

requirements of regulatory bodies are formally specified in the relevant standards documents.

Once a process has been mapped, the task of creating supporting process documentation is relatively straightforward. In the case of organizations that are pursuing registration to an industry QMS standard, the management representative should ensure that authors of the process documentation are provided the set of requirements in the QMS standard that apply to their respective processes and its documentation. The management representative also should ensure that there is a record kept of what QMS standard requirements have been allocated to the various personnel (and departments). This provides a mechanism to ensure that all requirements are allocated for implementation and none is accidentally ignored. Further, in cases where one requirement is allocated to more than one department (because one requirement can span multiple departments), such a matrix is helpful in identifying all parties that have been allocated a particular requirement and verifying that they are mutually consistent in implementing the particular requirement.

One frequently used method to keep track of the allocated requirements is to create a QMS traceability matrix that lists each QMS standard requirement (or group of related requirements; for example, requirements under one requirements clause or section) and traces each requirement to planned QMS elements (including QMS documentation) to satisfy the particular requirement. Such a QMS traceability matrix can be used to validate the QMS after implementation, that is, to determine whether or not each QMS requirement has indeed been satisfied (refer to Table 15 in Chapter 6). This traceability matrix also is useful for QMS maintenance; in the event of a change, such as to QMS documentation, it helps determine quickly what QMS requirements are applicable to that QMS document, and helps ensure continued compliance to applicable requirements. Conversely, if a QMS requirement is changed, the traceability matrix helps determine quickly what QMS elements are affected, and appropriate action can be initiated to maintain compliance to the modified requirements.

The set of QMS standard requirements, along with requirements from customers and internal stakeholders, might require changes to a process as it is currently executed. For example, a requirement might necessitate a change to the content of a document produced during a process, or it might require the insertion of a quality control activity in a process. It is the responsibility of the respective process owner to ensure that those authoring the process documentation have gathered applicable external and internal requirements, and that the process documentation is created and reviewed in accordance with the established process for creation of QMS documentation. (Refer to the section "QMS Documentation Process" in Chapter 3.) If an organization has a formally documented procedure that specifies the process for the creation of new procedures; the process owner must ensure that the procedure is adhered to. An example of such a procedure is provided in Appendix C.1.

The following is a general list of guidelines that should be followed when creating process documentation:

1. Enhance ease of use of process documentation by ensuring that documents of the same type have a standardized format.
2. Ensure that information in process documents is correct, clear, and unambiguous.
3. Ensure that different process documents are mutually consistent. For example, one procedure must not contradict another.
4. Ensure that the purpose, scope, and applicability of each process document are clearly defined.
5. Ensure that all process documents use consistent terminology.
6. Ensure that the roles and responsibilities associated with each process are specified clearly.
7. Ensure that the storage location of documents referenced in process documentation is clearly established and communicated throughout the organization.
8. Ensure that the format of process documentation is such that it can be used easily for employee training.

Besides familiarity with the process for the creation of process documentation, authors of process documentation should also be familiar with conditions that may necessitate changes to approved process documentation. Frequent changes to process documentation are generally discouraged, because it is harder to maintain and adhere to process documentation that is frequently revised. Maintenance problems arise from frequent revisions due to need for rereviews, reapproval, and redistribution of the revised documentation. Similarly, process adherence problems arise due to the need to frequently retrain employees on the revised documentation (or inform employees about the revisions). Employees find it harder to operate in a work environment where the rules are constantly changing, and such an environment invariably results in employee attitudinal problems such as indifference towards the QMS and low morale.

It is possible to avoid some of these problems by ensuring that the process documentation has the right amount of detail to preclude the need for frequent revisions. (Refer to related discussion in the section "Documentation Strategy" in Chapter 3.) Generally, the following conditions indicate when updates to process documentation might be required:

1. A change in requirements (external or internal)
2. A change in operational processes (including changes to address known deficiencies)
3. A need for further elaboration or clarification on information contained in the process documentation
4. An organizational change resulting in new roles and responsibilities (or changes to existing roles and responsibilities)
5. A change in tools and techniques used in the processes; for example, introduction of new tools, or elimination of tools currently in use

4.4 ESTABLISH A MECHANISM FOR PROCESS TAILORING (IF NECESSARY)

In certain industries, such as the software industry, product development companies with an established product development process often are faced with a real world problem: how can the defined process be used to provide adequate support for the execution of different types of product development projects in the organization? The crux of the problem is that a standard process, due to its generic nature, often requires that it be tailored to meet the requirements of each project in which it is used. A defined process may need to be customized or further elaborated to address the specific requirements (or characteristics) of a specific project effectively (see Figure 23). The "one process fits all" approach to product development projects does not always work, and differences in project environments often must be taken into account in the processes used.

The fundamental reason for the need for process tailoring is that no two projects are the same. Projects differ in terms of their characteristics, such as objectives, time schedule, budget, resources, priorities, and key drivers. All these factors have a profound effect on how a project is executed and what process is followed in executing the project. An organization may need to tailor its standard product development process to satisfy the needs and unique characteristics of each project. Failure to align the process in use with the project characteristics (i.e., lack of process tailoring) invariably results in inferior process support and a debilitating effect on the product development time, cost, and quality.

Process tailoring offers the following benefits:

Better project planning — Work to be performed in a project generally is decomposed into tasks and required deliverables. Planning estimates based on an improved awareness of the specific process to be executed, which is facilitated by process tailoring, are invariably better than those derived on the basis of a standard product development process.

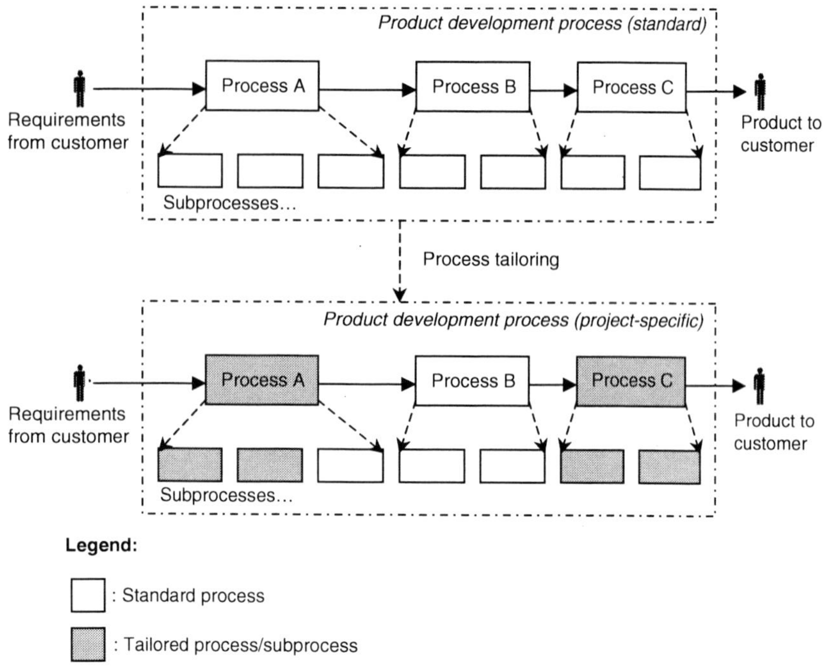

Figure 23 Concept of Process Tailoring

Improved process adherence — This results from fewer process deviation requests during project execution due to the proactive identification of deviations prior to start of execution.

Achievement of project goals — Process tailoring forces the project team to think about what is to be done and how to do it prior to actually starting to do the work. This helps minimize work disruption, wasted effort, and rework during project execution that would have resulted had such decisions been postponed until later in the project.

Deployment of process improvements — Deployment (or piloting) of process improvements planned for a project is simplified because they can easily be anchored in the project-specific process.

Increased predictability of outcome — An agreed-upon process helps minimize and eliminate any unknowns, increasing the probability that what is produced will be in line with the project team's expectations.

Enhanced auditing — Process tailoring supports the auditing process by proactively and explicitly identifying the process to be followed in a project. This makes the auditing process more effective because the auditor can audit the project against the project-specific product development process. On the other hand, if the auditor audits the project against the standard product development process, he is likely to discover discrepancies from the standard process during an audit because the same process cannot be applied "as is" across a set of projects due to reasons described earlier. Process tailoring facilitates proactive documentation of agreed deviations, along with supporting rationale by the project team. Thus, during a project audit, the auditor can focus on the adequacy and effectiveness of the tailored process.

As an example, a project manager may perform process tailoring by executing the following steps:

1. Elicit input from all departments involved in a product development project regarding the process to be followed.
2. Record planned major deviations against the standard product development process, along with supporting rationale and impact/risk assessment in the project plan (or project quality assurance plan).
3. Review the aforementioned planning document with a project team comprising management personnel from all departments involved in the project.

Once approved, the recorded deviations constitute the "informally tailored" process.

In addition to the concept of process tailoring, which in essence is a well-reasoned request for process deviation prior to project execution, there usually is also a need for a mechanism to request process deviations during project execution. The need for requesting process deviations during project execution might arise due to unforeseen project circumstances, such as schedule pressures, change in customer requirements, or other unexpected situations that necessitate a change in plans. In all such cases, when a process deviation is requested, it should be supported by a reason for the request, and adequate impact analysis and risk assessment performed by appropriate personnel, such as the submitter of the request, the project manager, and quality assurance department personnel. When appropriate, risk mitigation should be performed and a risk contingency plan should be created. Approved process deviations should be filed by the project manager (or quality assurance department personnel), and they should be monitored to determine whether or not associated risks remain acceptable; otherwise, activation of the risk contingency plan may be required.

5

QUALITY PRACTICES

As stated in Chapter 4, new process documentation must satisfy requirements of both internal stakeholders, customers, and external auditors. If a company chooses not to adhere to an industry QMS standard, its processes must satisfy quality requirements such that the defined processes are generally consistent with those prevalent in other product development companies. If a company wants to define an internal audits process, it first must ask what the key elements of the internal audits process of other companies are, specifically, the quality practices (or requirements) that generally apply to the internal audits process. It could then define a process to address those quality practices, and its internal audits process consequently would be generally consistent with those of other companies, and therefore acceptable and familiar to its customers.

This chapter provides a brief introduction to quality practices that are relevant to different functional areas in product development companies and therefore should be considered during QMS definition; it is not meant to provide an exhaustive list of requirements. This chapter can serve as a starting point for companies that are not pursuing implementation of any industry QMS standard, providing them sufficient guidance on the expected content of different product development processes. It is recommended that such companies also select one or more industry QMS standards as reference documents to identify additional requirements or quality practices for implementation.

An organization will realize immediate benefits by defining and controlling certain key processes. In certain cases, there might be dependency between processes that requires one process to be defined before another. Therefore, it is recommended that in the QMS definition phase, when appropriate, quality practices in the following subsections be assessed for criticality, anticipated benefit, and interprocess dependencies, and then be defined in the company in a prioritized fashion. The sequence of the quality practices in the following subsections is not intended to imply the sequence in which they should be addressed in this phase. Such prioritization also is necessary when personnel from one functional area are assigned the responsibility to define multiple processes, necessitating some sort of prioritization of the required processes to accomplish this task in a piecemeal manner.

5.1 TYPICAL PRODUCT DEVELOPMENT PROCESS

To examine the quality practices that pertain to different functional areas in product development companies, it is necessary to establish a context for this discussion. The overall product development process map provides this context and is shown in Figure 24. The product development process may be defined as:

A step-by-step process that results in the transformation of specified product requirements into a finished product by means of requirements specification, design, implementation, and testing.

The product development process comprises management processes, technical processes, and support processes:

Management processes deal with planning for product development, securing needed resources, management oversight for tracking and control of product development (including release authorization, when appropriate), and management of suppliers (if applicable).

Technical (or operative) processes are those that result in the transformation of process inputs (product requirements) into the final desired output (finished product for the customer).

Support processes, as the name suggests, are required to support the technical processes by providing the necessary organizational infrastructure (for example, workspace facilities and information technology infrastructure), or they pertain to ancillary business functions that are required in product development companies (for example, customer support process, employee training and continuing education process).

In Figure 24, dotted lines represent the alignment of the end of one process with the beginning of another when it is not otherwise obvious. For example, the alignment of the end of the requirements analysis process with the beginning of the product design process is clearly visible, so a dotted line is unnecessary. Such alignment indicates a dependency that requires one process to end before the other may begin.* For example,

* Typically, there is some overlap between dependent processes. For example, work on designing a product may begin before all requirements have been completely analyzed, as in the case of modular product construction, where requirements for certain product modules might have been completely analyzed and their design could begin before that of other modules. For the sake of simplicity, Figure 24 does not depict process overlapping.

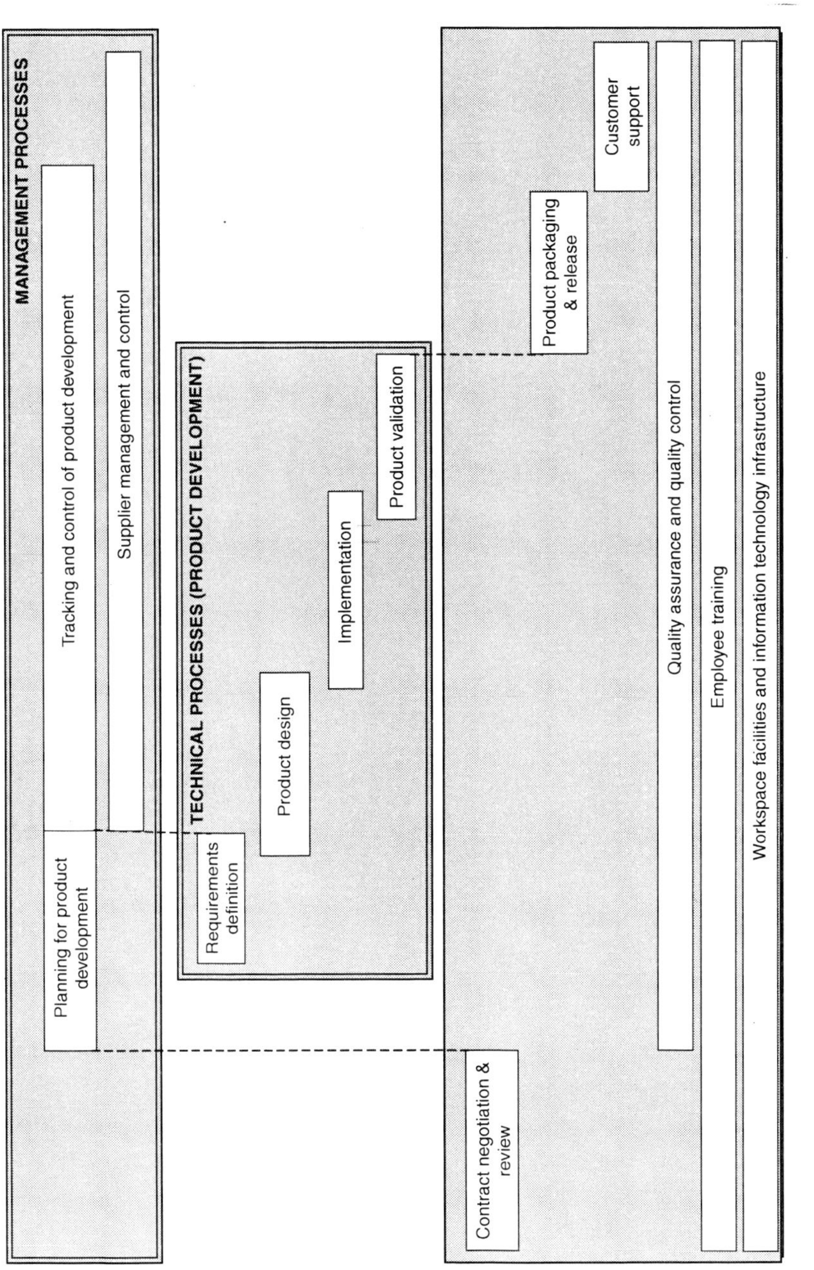

Figure 24 Product Development Process Map (including management, technical, and support processes)

assuming the company for which the product development process is depicted in Figure 24 is a contract-based company,* the end of the contract negotiation process marks the beginning of planning for product development and quality assurance for the new product. Similarly, with the end of product testing, work may begin to package the product and prepare it for delivery (subject to release authorization).

Figure 24 is meant only to show the three major types of processes and examples of processes belonging to each of these categories. It is not meant to provide a complete list of all functional areas or processes in product development companies (e.g., functions such as sales, marketing, and accounting). Other business processes that typically precede the processes shown in Figure 24, such as processes pertaining to identification of business opportunity** and definition of product concept,*** also are not depicted. The aforementioned functions and business processes are omitted from this discussion and from discussion on quality practices in the next section because they generally are outside the scope of industry QMS standards.

Note that the process map is for illustration purposes only, and the start and end time frame of each process is approximate. For example, the supplier management and control process is shown to begin once product requirements have been analyzed, and those designated for design and implementation by the supplier have been provided to the supplier. However, one could argue that this is an overly simplistic view of the process because supplier management and control (which includes supplier selection and qualification) begins much earlier. Finally, note that the process map does not end with the release of a finished product, but continues beyond that, and the length of each process does not necessarily correspond to its duration. For example, the customer support process

* A contract-based company is one that builds a product specifically to customer-specified requirements. A product-based company is one in which requirements are internally specified by taking into account market requirements that represent current and future needs of existing and potential customers.

** The business opportunity process encompasses the initial identification of a potential business opportunity from market research, or in response to customer input or requests for information (RFIs). The initial capture of a business opportunity is followed by preliminary senior management review to determine near- and long-term revenue potential of the opportunity, and a decision on whether or not to proceed with a detailed business case for the opportunity.

***The product concept process entails the high-level delineation of a proposed new product, or key enhancements to a current product, followed by a detailed senior management review of the business case for the product or enhancements. Authorization to proceed is contingent upon an acceptable business case. This assumes that the company is a product-based company, and not a contract-based company.

begins once a product is released to the market and continues until the company's support obligations remain in effect.

5.2 QUALITY PRACTICES IN PRODUCT DEVELOPMENT PROCESSES

The following subsections describe quality practices for various processes in product development companies. Each of these processes is categorized as a management process, a technical process, or a support process, as shown in Figure 24. Records for each of these quality practices should be retained when possible. Quality records are required for both internal and external audiences. They provide assurance to internal parties, such as management personnel, and to external parties, such as customers and regulatory bodies, that requirements for quality are being met. During process execution, process practitioners use quality records to verify that all requirements have been addressed and all required actions have been completed. Quality records also serve as evidence that can be audited in internal and external audits to verify that the processes are being executed in accordance with the defined QMS, and to identify opportunities for improvement.

5.2.1 Management Processes

As stated previously, visible and demonstrated senior management involvement and commitment to the QMS are essential for the effective deployment and use of a QMS in an organization. (Refer to the section on management commitment in Chapter 2.) Although management involvement in the development and continuous improvement of an organization's QMS is not a process *per se*, but rather an element of organizational culture, there are industry quality practices pertaining to management involvement and oversight of the QMS. We will address these before examining the quality practices that pertain to management processes.

Senior management has overall responsibility for the definition, deployment, maintenance, and continuous improvement of the QMS. Most product development companies establish a management body that exercises management oversight over the QMS and meets periodically, such as monthly or quarterly, to assess the adequacy and effectiveness of the QMS. This body includes senior management representation from all functional areas, is chaired by the quality management representative, and has a specified quorum (required minimum representation) for a meeting to be held.

Agenda items that should be addressed in the periodic management reviews include, but are not limited to:

- Review of open action items from past management reviews
- Establishment of short- and long-term quality objectives and monitoring of organizational performance against quality objectives
- Status of corrective actions
- Status of preventive actions
- Status of capability improvement initiatives
- Measurement results and trends
- Results of internal and external audits and assessments
- Customer satisfaction data
- Supplier performance
- Records of periodic management reviews of the QMS, such as the agenda for each meeting, presentation material, and meeting minutes, should be stored in a secure location.

In addition to management involvement in the form of periodic management reviews of the QMS, senior management should demonstrate its commitment to the QMS by adequately discharging the responsibilities described in Chapter 2 under the discussion on the QMS Steering Group. Senior management also should appoint a member of senior management as a quality management representative with day-to-day responsibility for overseeing the definition, deployment, monitoring, and continuous improvement of the QMS. (Refer to Chapter 2 for desired qualifications of a quality management representative.)

5.2.1.1 Planning for Product Development

Planning for the design and development of a product entails the specification of information such as design and development stages, along with associated milestone dates, required resources, and responsibilities. Most organizations designate a project manager who has overall responsibility for planning, tracking, controlling, and coordinating project execution. To ensure consistency in planning for product development, there should be an established process, and product development plans should be based on standardized product development plan templates that have been reviewed and approved. Planning for quality assurance and quality control tasks in a project generally is covered in a separate project quality plan (discussed later in this section).

A standardized template for a product development plan should cover items such as:

- Purpose and scope of the product development project
- Specification of project organization structure, and allocation of responsibilities for the project (including role and responsibilities of suppliers, if any)

- Development process to be followed, including planned project phases and key milestone dates for each phase
- List of planned product deliverables, including documents and all artifacts to be produced (including product prototypes)
- Reference to other project planning documents, such as the project quality plan
- Estimates for the project, including:
  1. Resource requirements:
     - Cost (including labor, travel, and material cost)
     - Personnel
     - Required equipment
  2. Detailed project schedule listing project activities and deliverables. In creating the schedule, a detailed work breakdown structure for the project should be created, and a critical path analysis should be performed.

     Estimates should be based on historical data from similar projects and, when appropriate, derived using formal estimation techniques. Appropriate amount of contingency should be included in the estimates. The basis of estimates should be documented so it can be explained how each estimate was arrived at.
- Customer and supplier involvement during the project, such as joint reviews and approvals
- Assumptions and dependencies associated with the project
- Initial identification of project risks, and risk mitigations and contingencies. It generally is preferable to maintain a separate project risk management plan, similar to the one shown in Table 8.
- Employee training needs due to unique requirements of the project

Insofar as quality planning is concerned, the organization must first identify the processes (including quality control activities), documents, and resources it will need for producing a high-quality product. Typically, the product development process and its subprocesses, including description of required resources and templates and forms for documents to be created during the processes, are established as standardized QMS documentation. Each project* must adhere to the applicable QMS documentation. Quality planning should address the need to tailor standardized QMS documentation for a specific project. The output of quality planning can be a stand-alone project quality plan (usually created by the quality assurance department), or the information may be distributed in one or more project planning documents.

* Here, a project also can refer to a product or contract.

Information that should be included in a project quality plan includes but is not limited to:

- Purpose and scope of the project quality plan
- References to related project planning documents
- References to relevant QMS documentation applicable to the project
- Specification of, or reference to, quality requirements applicable to the product
- Specification of, or reference to, applicable product release criteria
- Specification of, or reference to, how failure to comply with any release criteria is to be handled
- Specification of, or reference to, entry and exit criteria for each project phase
- List of planned verification, validation, monitoring, inspection, and test activities
- List of project metrics
- List of process exemptions* (or process deviations) from the QMS documentation approved in advance for the project (along with supporting rationale and risk mitigation). Note that management personnel from all functional areas involved in the project provide this information to the author of the project quality plan.

5.2.1.2 Tracking and Control of Product Development

Once product development gets underway, project progress should be monitored closely. Project execution should be monitored to verify that project execution is proceeding in accordance with established plans, and to implement timely corrective and preventive actions, when necessary. Project tracking and control should include activities such as:

- Formal periodic project review meetings with management personnel from all relevant departments in order to determine status (including slippage from the schedule) and to identify and address problems, resource needs, and project risks (refer to Appendix D.4 for an example of a template to record minutes of project review meeting)
- Day-to-day project tracking and control, including but not limited to:

* Also, if preferable, instead of listing them in the project quality plan, exemptions can be listed in the appropriate project plans. For example, deviations from the standard product test process may be noted in the product test plan.

  - Planning for day-to-day tasks (tactical planning and replanning)
  - Prioritization of tasks
  - Resolution of interdepartmental conflicts
  - Follow-up on open issues and action items
  - Risk management
  - Other such tasks necessary to ensure project success
- Formal milestone review at each critical project milestone, to determine whether or not all objectives (criteria) associated with that milestone have been met, and to determine necessary actions when objectives have not been met (refer to Appendix D.5 for an example of a project milestone review report form)
- Monitoring of expenditure against budget in order to control expenses and request additional funds, if necessary
- Monitoring of resource utilization in order to reallocate resources or request additional resources, if necessary

When necessary, the original product development plan should be revised and subsequent phases of the project replanned if it is apparent that adherence to original plans is not feasible.

5.2.1.3 Supplier Management and Control

A supplier provides a product or service to an organization. The supplied product can be included as part of the finished product delivered by the organization or used during the product development process.

An organization should select its suppliers after a careful evaluation of each supplier's ability to meet current and future organizational requirements and expectations. The extent of the due diligence an organization needs to perform in the supplier selection process and the extent of the subsequent supplier control will vary with the perceived criticality and impact of the supplier to the finished product and/or to the overall product development process.

Sometimes organizations that are registered to an industry QMS standard limit their supplier selection process to a single requirement: that the new supplier be registered to the same quality management system standard. This provides the organization a high degree of assurance that the supplier will operate in the same quality framework, because it will be governed by the same standard. However, for various business reasons, it is not always advisable to restrict supplier selection in this manner. That is, registration to a particular industry QMS standard might be desirable but should not be necessary. Regardless of whether a potential supplier has the desired quality registration, it is always preferable for an organization to determine the quality of its potential supplier by performing an assessment itself. Due to the subjective nature of auditing, and the fact that no two registrars have exactly the same interpretation or implementation expectations for each requirement in the applicable quality management system standard, there are bound to be differences in the quality of QMS implemented by two different suppliers registered to the same industry QMS standard. While the two suppliers may have a QMS in place, the QMS of one supplier may be distinctly superior (in terms of effectiveness and maturity). An example of a recommended process for supplier selection is provided in the supplier management and control procedure in Appendix C.3.

The supplier management and control process should address the following items:

- An evaluation of the capability of each potential supplier to supply product that meets specified requirements, including technical requirements, quality requirements, and business requirements (for example, cost and timeliness)
- Clearly defined criteria that must be met for the selection of any new supplier
- Mechanisms for continually monitoring, evaluating, and controlling supplier performance
- Identification of means and/or documents, such as a legal contract, that will be used by the organization to communicate and formalize agreement with its suppliers on purchasing requirements, and mutual expectations and obligations
- Identification of conditions that might cause a supplier to be removed from an organization's list of approved suppliers
- Verification of the purchased product against requirements to determine whether or not specified requirements were met
- Conditions that must be met prior to release of the purchased product for use within the organization
- Mechanisms that specify how a purchased product that does not meet specified purchasing requirements will be handled
- Means to identify uniquely each purchased product in the product development process, and in the finished product delivered by the organization

5.2.2 Technical Processes

Technical processes for product development encompass activities pertaining to product requirements definition, product design, implementation, and product validation. The following subsections examine quality practices that are applicable to each of these technical processes.

5.2.2.1 Requirements Definition

The technical process of product development begins with the definition of product requirements. The requirements definition process essentially comprises three steps:

1. *Requirements determination* (or *elicitation*): first, product requirements are elicited from all appropriate requirements sources;
2. *Requirements documentation*: second, the identified requirements are formally documented; and
3. *Requirements review and approval*: finally, the documented requirements are reviewed by all stakeholders and formally approved by relevant parties*

Possible sources of product requirements that should be considered during requirements elicitation include but are not limited to:

- Customer requirements, including:
  - Requirements explicitly specified by the customer
  - Implicit customer requirements — that is, requirements not specified by the customer but essential to support an explicit customer requirement
- Requirements internally specified by the organization, including:
  - Requirements explicitly specified by the organization — this is usually in the case of product-based companies; in case of contract-based companies, such proposed requirements should be reviewed and approved by the customer
  - Implicit requirements that are necessary to support the implementation of an explicit requirement specified by the organization or an otherwise known use of the product
- Requirements due to applicable statutory and regulatory bodies

When documenting product requirements, care should be taken to ensure that the requirements satisfy the following attributes:

1. *Clarity* — that is, each requirement should be unambiguously worded.
2. *Completeness* — that is, requirements should not be missing necessary information, and necessary requirements should not be missing altogether.

* Guidance on how to identify relevant parties for the approval of a document was provided in Chapter 3.

3. *Atomicity* — that is, each requirement should be uniquely identifiable (and therefore traceable during product design, development, and test).
4. *Correctness* — that is, each requirement should be technically sound.
5. *Testability* — that is, it should be possible to validate either by product testing, inspection, or some other means whether or not each requirement has been satisfied.
6. *Feasibility* — that is, the requirements should be achievable within the parameters of schedule, cost, and resources (including technical ability of resources to meet the requirements).

Once product requirements have been documented, they should be reviewed by all stakeholders to verify that the documented requirements satisfy the aforementioned criteria. Records of such requirements reviews should be created and maintained for future reference, along with a record of how the action items from the review were handled. Requirements review should be performed prior to making a commitment to the customer (or other parties within the organization) for the supply of the product. Requirements should be formally approved only after they have been reviewed and reworked to the satisfaction of all stakeholders (and the customer, where applicable).

Once product requirements have been approved, they may need revision to reflect changes requested by the customer, to address a need for clarification or correction identified during product design and development, or for other unavoidable reasons. For this reason, it should be clearly identifiable and known:

- Who may authorize requirements changes
- How changes to previously approved requirements are clearly identified
- The impact, if any, of a requirements change on other product requirements or on parts of the product already designed or implemented
- Who reviews and approves requirements changes
- How all affected personnel in the organization are notified of requirements changes

Note that, in the case of contract-based companies, a change in key requirements may necessitate a contract amendment.

Finally, beginning with requirements definition and concluding with product testing, there should be means to trace design, implementation, and testing of each product requirement. Such a traceability mechanism is necessary to ensure that each product requirement is designed into the product, is implemented, and is tested to verify that it was met.

5.2.2.2 Product Design

This process begins with requirements analysis and ends with a detailed specification of the product design that satisfies product requirements. Following are the key steps in the product design process:

Analysis of product requirements — In this step, product requirements should be carefully analyzed to gain a precise understanding of the requested product capabilities.

High-level design (design outline) — Once product requirements have been analyzed, product architects should create a high-level design outline wherein the product is decomposed into key components. The design should be represented by drawings, specifications, or any other form that permits verification of the design against product requirements. Product requirements then should be allocated to appropriate components to facilitate low-level design of each component. An organization might not design and develop all components of a product; it might decide to outsource some or all of these activities for one or more components. Such build-versus-buy decisions should be finalized at this time (if not done previously).

High-level design review — The high-level product design should be reviewed with all stakeholders for correctness, completeness (design should satisfy all applicable product requirements), acceptability (by relevant stakeholders), feasibility, inherent risks, and other considerations. The high-level design should be reworked as required and approved prior to creation of low level of design.

Low-level design — Once the high-level design is approved, each component should be designed in accordance with the product requirements allocated to it. The low-level design should be presented in a form that permits verification against product requirements and contains sufficient information for implementation (including production by a supplier, if applicable). If the need for a prototype of the product (or for parts of it) was identified in the product development plan (or is formally required as per product requirements), the prototype should be developed at this stage.

Low-level design review — The low-level design should be reviewed with all stakeholders for considerations such as those described under high-level design review. The low-level design should be reworked as required and approved prior to commencement of design implementation activities. This includes verification of the product prototype by appropriate means, such as testing, and redesign and refinement of the prototype, as needed.

Records of high- and low-level design reviews should be created and maintained for future reference, along with a record of how the action items from the review were handled. In case of product prototypes, this entails maintaining records of the verification of the prototype and additional actions taken.

Product design, once approved, may need to be modified to address deficiencies identified later in the product development process, to satisfy new or modified requirements, or for other reasons. Issues that should be addressed in the event of a design change are similar to those described for requirements changes in the previous section.

5.2.2.3 Implementation

Implementation may begin after the product design is approved. During this step, the product should be implemented in accordance with the approved design, by following applicable workmanship standards. Workmanship standards should be created to facilitate consistency and to minimize injection of defects during product development. When applicable, relevant industry workmanship standards should be used.

This process results in the final product or in components of the final product. During implementation, the product (or components) in development should be verified periodically by means of testing, inspection, or other appropriate means, in order to assess conformance to product design and to identify product defects. Records of such verification activities should be created and maintained for future reference, along with a record of actions taken. The product should be reworked (and reverified) as needed to comply with the design, and for the elimination of known product defects. When necessary, the nonconforming or defective product should be scrapped and reimplemented. If the design is deemed deficient, the process for requesting changes to the approved product design should be initiated. Once known nonconformities or defects in the product have been eliminated, the product may be released for validation testing.

Tools used for product verification or validation (which is described next) should be calibrated periodically to assure accuracy of verification and validation results. The calibration status, including date of last calibration, should be identified clearly. Tools should be protected from tampering that would affect their calibration status and compromise verification and validation results.

Finally, mechanisms should be available to identify and trace a product and its components during the product development process.* Identification and traceability mechanisms are necessary to track each product requirement during design, development, and testing, to identify and isolate defective products or components; to trace requirements and design

* In manufacturing, this entails traceability using manufacturing dates, production lots, and product serial numbers.

changes to affected products and components; and to track the status of the product and components with respect to the product testing process. After product release, such identification and traceability mechanisms are invaluable for identifying products being recalled or needing to be replaced or modified.

5.2.2.4 Product Validation (Testing)

Once a product has been developed, it should be validated against requirements to determine whether or not all product requirements have been satisfied. Product validation typically is performed by means of testing of the product against a defined test plan. Test results should be recorded and compared against expected results to determine if the tests are successful. In case of failure, identified defects should be logged formally and corrected by following a defined defect management process. The defective product should be controlled to preclude its delivery to the customer or other unintended use in the product development process. Once product defects have been corrected, product testing should be repeated to verify that previously discovered defects no longer exist. The results of testing, including defects reported, should be maintained for future reference.

When it is not feasible to validate a specific use of the product by means of testing, all such test limitations should be clarified up front, and risks associated with these limitations should be identified and mitigated. If certain requirements cannot be tested, the product can be inspected to confirm that the requirements were met. In such a case, or where a product with known nonconformities or defects needs to be released to the customer prior to elimination of the deficiencies, it may be released under concession from relevant management personnel in the organization or from the customer (when appropriate) after consideration of associated risks and mitigation of the same.

5.2.3 Support Processes

5.2.3.1 Contract Negotiation and Review

This process entails documenting the terms and conditions of a product sale in a legally binding agreement between the buyer and the seller. Prior to contract approval, the buyer and seller negotiate and review the contract until the terms and conditions are acceptable to both parties.

The following are some of the items that should be addressed during definition of this process:

- Ensure that the contractual agreement accurately reflects the current and/or planned features and capabilities of the product so as not to preclude the satisfaction of any contractually agreed items.
- Ensure that each contractual requirement can be validated to confirm it was met.
- Ensure that the requirements are complete and there are no missing requirements.
- Ensure that the requirements are clearly and unambiguously documented.
- Ensure that the contractual requirements are achievable within the parameters of schedule, cost, and resources (including technical ability of resources to meet the requirements). For this purpose, the draft contractual agreement should be reviewed with relevant personnel within the organization.
- If a product is being built to contract, and periodic joint reviews of project progress are required, clarify when such reviews will take place, or clarify the mechanism for scheduling such joint reviews.
- Confirm standards or specific processes that are contractually required to be followed during product development.
- Clarify use of customer-supplied items, if any, including criteria for the acceptability of customer-supplied items, and the mechanism for handling those deemed unsuitable for use.
- Review responsibilities of the customer under the contract.
- Ensure appropriate nondisclosure agreements are in place to protect the intellectual property of the organization and the customer.
- If applicable, agree on requirements pertaining to safekeeping of a true copy of the product in an escrow location (for use in the event of a disaster).
- Communicate warranties associated with the product.
- Confirm the duration of support requested by the customer (standard support or extended support).
- If the customer has requested to perform quality audits of the organization, agree upon when, where, and how such audits may be scheduled and performed.
- Ensure that the signed copy of the contract is stored in a secure location, such as a fireproof location with access privileges to authorized personnel only.
- Agree on a mechanism for implementing future amendments to the contract, and for informing relevant personnel in the organization in the event of a contractual amendment.
- Finally, records of contract reviews — for example, in the form of minutes of meetings — should be maintained, along with a record of how the action items from the reviews were handled.

5.2.3.2 *Product Packaging and Delivery*

This process deals with the packaging, handling, storage, and delivery of a finished product to the customer so that the product is protected from damage and tampering, and is handled by authorized personnel at all times.

Items that should be addressed during definition of this process include but are not limited to:

- Packaging specifications containing detailed instructions should be documented so that different personnel package and label the product in a consistent manner.
- Material and equipment required for packaging and handling the product should be made available.
- Packaging and handling equipment should undergo scheduled inspection and maintenance. When appropriate, the date of last scheduled inspection and/or maintenance, the name of the person who performed the task, and the date of the next scheduled inspection and/or maintenance should be identified clearly on the equipment.
- Methods for handling the product should be defined, to protect the product from damage.
- Personnel packaging and handling the product should receive appropriate training and equipment, and records of employee training should be maintained.
- There should be clearly designated areas for storing packaged products until delivery to the customer. Packaged products should remain segregated until they are shipped to the customers.
- Packaged products should be uniquely identified so that they may be quickly isolated, as for example in the event of a product recall, or to aid traceability in the event of a product defect.
- When applicable, shelf life markings should be displayed clearly on product packaging.
- The movement or transfer of products from storage areas should be controlled, and appropriate authorities for controlling receipt and release of the products should be defined.
- The inventory of packaged products should be monitored to track inventory level changes due to receipt of products from production and release of products to customers.
- The inventory of packaged products should be audited periodically (e.g., monthly) to verify that stored products are not damaged, deteriorated, lost, or otherwise unfit for delivery to the customer. For products with expiration dates, the audit should verify that the expiration date of products has not expired, or the products should

be promptly disposed. It also should be verified that products are arranged in the inventory such that products with earlier expiration dates are located before products with later expiration dates.

- Products should be delivered by the means that were contractually agreed upon with the customer (including use of contractually required packaging material).

5.2.3.3 Customer Support (including Customer Training)

The customer support process (also called the customer service or customer care process) constitutes the maintenance phase of the product. It covers all aspects of customer service after delivery of the product to the customer. This includes handling and processing of customer calls pertaining to:

1. *Customer-reported problem* — this includes requests for product replacement, repair, parts, or service under product warranty or service contract.
2. *Customer complaint* — this is a report of an unpleasant customer experience, request for process change due to customer perceived deficiency, and so on.
3. *Request for information* — this includes requests for additional information on how to set up a product, how to operate it, clarification on information provided in product documentation, information on future enhancements, and so on.
4. *Customer suggestion* — this includes requests for product enhancements or process improvement.

Items that should be addressed during definition of the customer support processes include but are not limited to:

- Definition of the process for resolution of customer-reported problems and complaints, including:
  - Customer-reported problems and complaints should be resolved in a timely manner, commensurate with the nature and severity* of each reported problem. The response time should be in accordance with the prescribed** response time for a customer-reported problem (or complaint) of that severity.

---

* The severity of a customer-reported problem generally is agreed upon by the organization and its customer.

** Most companies generally establish prescribed response times for customer-reported problems and complaints of different severities. Each instance of a reported problem or complaint then is required to adhere to the prescribed response times.

- It should be determined whether the customer-reported problem or complaint pertains to improper use of the product by the customer, a product defect, or an observed deficiency in one or more of the organization's processes. In case of improper use of the product by the customer, necessary information should be provided to the customer to use the product correctly. With a product defect or process problem, the root cause for the reported problem should be systematically derived and addressed. In the case of a product defect, this may entail replicating the customer-reported defect.
- When possible, the proposed solution for a product defect should be verified prior to providing the same to the customer. A proposed solution might include product replacement or repair, replacement of the defective parts, or onsite service, as appropriate. Until the reported defect can be resolved, the customer should be provided temporary relief by means of a workaround (temporary solution), if possible.
- When a reported problem or complaint cannot be resolved immediately and requires several days for corrective action, then the customer should be kept informed regarding the status of the request and the anticipated closure time.
- Records of customer interaction related to the reported problem or complaint and records of corrective actions taken should be maintained.
- It should be verified with the customer that the provided solution has resolved the problem or complaint.
- Customer satisfaction data for the entire experience, from the reporting of the problem or complaint to problem resolution, should be collected and used for process and product improvement.
- An examination should be made of how such problems and complaints may be prevented in the future — for example, by improving product documentation, or by requesting an appropriate change to the relevant product development or test process.

■ Definition of the process for handling requests for information from the customer, including such issues as:
  - Requests for information should be forwarded to the department or personnel most suitable for responding to the request.
  - Prior to release of the customer-requested information, it should be ensured that authorization from appropriate personnel is obtained for release of the information, especially in the case of company confidential information.

- When a request for information cannot be fulfilled immediately and requires several days to be addressed, the customer should be kept informed regarding the status of the request and the anticipated closure time.
- Records of customer interaction related to the request for information and actual information provided should be maintained.
- An examination should be made of how such customer requests for information may be prevented in the future (for example, by making necessary changes to product documentation).

■ Definition of the process for handling customer suggestions, including such issues as:
  - Review of customer suggestions should include personnel from the departments most suitable for considering and deciding on the disposition of the suggestions. For example, a customer suggestion that impacts the work performed by a specific department should be reviewed with personnel from that department, as well as other appropriate departments, such as quality assurance.
  - Customer suggestions should be addressed in a timely manner, commensurate with the nature and severity* of each suggestion.
  - The customer should be kept informed regarding the status of the customer suggestion and the anticipated closure time.
  - Records of customer interaction related to the customer suggestion and final disposition should be maintained.
  - The customer who made the suggestion should be informed about the disposition of the request, including reason for rejection, when applicable.

For all the aforementioned types of customer calls, customers should be provided information regarding how to contact the organization for customer support. Processes for handling different types of customer calls should address handling of customer calls during and outside of regular business hours. For example, customers who have purchased premium support may be entitled to customer support outside of regular business hours. Further, the process should define the expected time frames and methods for responding (including initial acknowledgement) to each type of customer call. This includes escalation mechanisms within the organization in case prescribed response times are not met.

The customer training process entails implementation and maintenance of a customer training program covering all aspects related to training a

* The severity of a customer suggestion generally is determined internally by the organization.

customer and the organization's own employees on the organization's products. Items that should be addressed during definition of this process include but are not limited to:

- Identification of customer training courses based on customer needs analysis
- Specification of training tracks (list of required and recommended courses) for each type of training audience
- Identification of course objectives and course content for each training course
- Determination of logistics associated with each course (e.g., duration, frequency, cost, audience, delivery method*, required equipment, and training material)
- Process for developing new training course material, including piloting of the course material (when necessary)
- Process for training administration, including notification of available courses and registration of students
- Method for measuring and using student evaluation** — that is, change in student understanding and competency from before and after the training to determine whether or not course objectives were met, and for improving the course material, as needed.
- Method for collecting, implementing, and tracking student comments and suggestions regarding the course

5.2.3.4 Quality Assurance and Quality Control

Although quality assurance and quality control activities do not collectively form a distinct process in an organization, they are represented as one here for the purpose of discussing quality practices that apply to them. This is required because the extent of quality assurance and quality control has a direct bearing on the quality of the delivered product.

Quality assurance and quality control activities generally are deployed throughout an organization's QMS. (Refer to the explanation in Chapter 1.) Therefore, they are executed by personnel from different departments in an organization, although one department — the quality assurance department — typically exercises oversight over all quality assurance and quality control tasks.

---

* Examples of training delivery methods include Web-based training (WBT), computer-based training (CBT), classroom study, self-study, on-the-job training, conferences and seminars, tutorials, mentoring, and continuing education at college or university.

** Student evaluation is usually performed by administering a quiz on the subject matter before and after the training, and using the change in student evaluation scores to assess effectiveness of training material and identify areas for improvement.

Items that should be addressed during definition of quality assurance and quality control activities include but are not limited to:

- A QMS should be defined to represent the organization's business processes (and their interaction) that result in the delivery of the final product or service to the customer.
- Business processes should be defined, and documented to the extent necessary for consistent process execution. Definition of business processes should address required resources, key activities, personnel responsible for the various activities, supporting methods and tools necessary for effective process execution, and criteria and requirements to be satisfied by the process. (Refer to Appendix D.1: Procedure Template.)
- A plan, typically the project quality plan, should exist for the achievement of quality requirements in a product development project. (Refer to guidance provided earlier in this chapter regarding expected content of a project quality plan.)
- Quality records should be identified to provide evidence of conformance to requirements and of results achieved. Note that quality records are required not only to provide assurance to internal and external parties that quality requirements are being met, but also are used in management decision making, such as whether or not to release a product to customers. Quality records should be clearly identifiable, complete, stored in a secure location, retained for an appropriate period of time, and accessible to relevant personnel.
- Internal quality audits should be performed periodically to assess adequacy and effectiveness of the QMS. Internal audits are discussed in detail in Chapter 7.
- Means such as process measurements should be planned and used to monitor process execution for the purpose of statistical process control (process execution within permissible bounds), to monitor process performance against established quality objectives (or criteria), to determine if process improvement initiatives are resulting in improved process performance, and so on.
- Quality control activities should be established across the organization to compare actual process output against required process output. Take appropriate corrective action, when necessary.
- Active risk management should be performed during product development to monitor, control, and mitigate quality-related risks. This entails identifying quality-related risks, devising a risk mitigation and contingency plan for each known risk, and actively monitoring the risks to determine need for activation of the risk mitigation and contingency plans.

- If some of the organization's processes are outsourced, or the organization obtains parts for use in the final product, the QMS should include the means used to monitor and control supplier performance.

5.2.3.5 Employee Training

This process deals with identifying training needs for employees, providing the training, and maintaining employee training records. This includes product training, technical training, and soft-skills training. Items that should be addressed during this process are discussed in Chapter 7.

5.2.3.6 Workspace Facilities and Information Technology Infrastructure

As with quality assurance and quality control, workspace facilities and information technology (IT) infrastructure are not a distinct process *per se*, but encompass activities essential to providing a suitable environment for product development and support. These activities provide the physical infrastructure, including IT equipment, necessary for the execution of product development and support processes. This includes but is not limited to:

- Physical workspace — that is, buildings, rooms, storage spaces, and so on
- Physical work environment — cleanliness, safety, lighting, protective equipment, pollution, heat, humidity, and so on
- Transportation equipment
- Machinery
- Packaging material
- Office automation equipment such as fax machines, printers, copiers, phones, computers, servers, routers, databases, operating systems, and office automation software

Items that should be addressed during definition of these activities include but are not limited to:

- Planning of an organization's processes and planning for product development should include an identification of (and request for) required workspace facilities and IT infrastructure.
- During process execution, it should be ensured that identified workspace facilities and IT infrastructure are available. The responsibilities pertaining to these tasks should be identified clearly.

- Workspace facilities and IT equipment should be maintained; this includes periodic inspection and recertification (where appropriate), and ability to address requests for repair and maintenance in a timely manner.
- It should be ensured that an organization's workspace facilities and IT infrastructure are protected from (and can be recovered from) disaster scenarios. Any scenario that causes a prolonged business interruption may be viewed as a disaster that requires advance planning for contingency measures. Examples include limited or significant damage to a facility due to environmental catastrophes or accidents, power outage, terrorist strike, and security breach of a facility. The disaster recovery plan should describe the recovery process, roles and responsibilities, and schedules to minimize the impact of these scenarios and facilitate a rapid recovery.

PHASE III

REFINE

6

QUALITY MANAGEMENT SYSTEM REFINEMENT

Once a QMS has been defined, QMS implementation progresses to the next phase: QMS refinement. The three main objectives of this phase are:

1. Verify that all QMS documents are mutually consistent with respect to their interfaces, terminology, format, and content.
2. Validate that all QMS documents collectively satisfy all applicable QMS requirements.
3. Take necessary corrective action to address the deficiencies identified during this phase.

While the focus in the QMS definition phase is on defining, verifying, and validating *individual* elements of the QMS, the focus in this phase shifts to verifying and validating the *complete* system. This provides a high degree of assurance that all QMS requirements have been adequately addressed across the entire system, and none has fallen through the cracks.

The following subsections describe in detail how the above objectives can be achieved.

6.1 VERIFICATION OF COMPLETE QMS

Verification of the defined QMS is best accomplished by having two or more quality assurance department personnel independently review all Tier 1, 2, and 3 QMS documents. Such a review may be limited to the quality manual, procedures, and work instructions only. This review can only be effectively performed once all the required QMS documentation has been created; therefore, this review should not be initiated until the QMS definition phase is complete. If an organization performs a phased rollout of its QMS, this review should be performed once all planned

QMS documentation has been documented. This implies that in a phased rollout, QMS documentation — once created and approved — is followed by deployment of that QMS documentation, while the refinement of that QMS documentation has to wait until all planned QMS documentation has been created. All QMS documentation is then verified; this may result in changes to QMS documentation that has not yet been deployed, as well as changes to documentation that has.

The following is a checklist of some of the items that should be verified during this review. An organization may add additional items to this checklist as per its unique needs.

- Is there content consistency between QMS documents? Ensure that the information contained in one QMS document is not contradicted by another.
- Is there consistency in terminology between various QMS documents? Specifically, ensure that all QMS documents use standard terms and acronyms that are defined in one place and accepted throughout the organization. Authors of QMS documents should refer to one QMS document that contains the approved definitions of all QMS-related terms. Document authors should not redefine QMS terms in their respective documents; that results in terms having multiple meanings, which in turn causes confusion and misunderstanding regarding what is supposed to be standardized QMS terminology.
- Is there format consistency between QMS documents of the same type? Inconsistency between documents of the same type can result from the template for the document evolving over time. For example, an organization might have documented some of its QMS procedures by using a procedure template that was subsequently revised, and later might have used the revised template to document other QMS procedures. In such a case, it would be desirable to migrate the procedures from the old template to the new template, if the change is significant and migration cost is not prohibitive.
- Is there referential accuracy between QMS documents? If a procedure 'x' states that certain information is documented in procedure 'y', verify that procedure 'y' contains that information.
- Is there interface handshaking between the various QMS documents? If procedure 'x' states that it receives an input 'xx' from the procedure 'y', verify that procedure 'y' identifies 'xx' as one of its outputs.
- Is there redundancy between QMS documents? Ensure that a process (or parts of it) is not described in more than one document. Information redundancy creates document maintenance problems and introduces the risk that different documents, which may be consistent initially, may become inconsistent over time.

The reviewers can summarize their results using a simple spreadsheet where each row lists a QMS document along with the reviewer's comments. All the reviewers should meet and should jointly review the comments made about each QMS document. Once there is agreement on needed changes, document change requests should be initiated in accordance with the document change request process described in Chapter 3.

6.2 VALIDATION OF COMPLETE QMS

For companies implementing a QMS in accordance with a quality management system standard, QMS requirements are specified in the standard and in the company's QMS documentation. For companies that are not using a quality management system standard, all QMS requirements from internal stakeholders and customers are specified in the company's QMS documentation only. Validation of the complete QMS is strongly recommended in case of companies that are implementing a QMS in accordance with a quality management system standard, although companies that are not using a quality management system standard also can perform such a validation by listing internal quality requirements in lieu of quality management system standard requirements. The purpose of this validation exercise is to perform a gap analysis to identify quality requirements that have not been addressed, due to oversight, piecemeal implementation* of one or more quality requirements, or communication gaps between various departments and process owners involved in the implementation.

Validation of the complete QMS entails the following steps:

1. Identification of all applicable requirements; and
2. For each quality requirement, reference to *auditable* evidence** that demonstrates that the requirement has been satisfied in the QMS; for example, reference to a procedure that describes the execution of a required process.

* A quality requirement is implemented in a piecemeal fashion when the requirement is split into pieces, and implementation of the pieces is spread over a period of time, or across different departments.

** Most quality management system standards do not require that conformance to each requirement be demonstrated by documenting how the required process or activity is executed in the company. In other words, certain requirements may be satisfied simply by executing the required process or activity. Adherence to applicable requirements can be verified in an audit by observing the execution of the process or activity, or by interviewing relevant personnel and examining outputs of the process or activity. As an example, the year 2000 version of the ISO 9001 standard (ISO 9001:2000) requires that only 6 processes be formally documented, compared with 18 processes required as per the year 1994 version of the standard (ISO 9001:1994).

The task of performing this validation exercise is considerably simplified if one uses the QMS traceability matrix described in Section 4.3, Chapter 4, and shown in Table 15. It should be verified that the elements traced do indeed demonstrate complete conformance to the applicable requirement, and deficiencies (gaps) should be noted. This examination may cause the traceability of some requirements to QMS elements to change from what was originally envisioned when the requirements were allocated for implementation. For example, it may be determined that a requirement needs to be traced to additional QMS elements. Gaps identified during this exercise should be addressed in this phase by requesting necessary action from the respective process (or document) owners.

Table 15 QMS Traceability Matrix

Requirement Statement (or Identifier)	*QMS Element*	*Process/Document Owner*	*Compliance (Check One)*			*Gap Statement*	*Assigned To:*	*Due Date*
			Full	*Partial*	*Not Compliant*			
(Record the requirement statement or a unique identifier for the requirement (if one is available). When necessary, the source of the requirement should be noted, such as quality management system standard, or internal company policy.)	(Identify the QMS element that demonstrates adherence to this requirement. The QMS element may be a reference to a QMS document, activity, process, or other auditable evidence.)	(Identify the department [or person] that owns the referenced QMS element. This identification is to be used to request corrective action in case a gap is identified in the implementation of this requirement.)				(Provide a precise statement of the noted gaps in adherence to the requirement.)	(Identify a person in the quality assurance department who will work with the process/document owner to close the noted gap.)	(Identify the date by which the gap is required to be closed.)

PHASE IV

DEPLOY

7

QMS DEPLOYMENT

QMS deployment entails the promulgation and institutionalization of the defined QMS of an organization. Specifically, it means that the way an organization executes its business processes is consistent with its QMS documentation.

QMS deployment is essentially a three-step process:

1. **Employee training** — Employees are provided QMS training relevant to their jobs and are then required to adhere to applicable QMS processes and procedures. QMS training is not a one-time event, nor is it performed only once the complete QMS has been defined. In fact, employee training on the QMS is ongoing because:
   - QMS definition is typically piecemeal, wherein parts of the QMS are gradually defined and incrementally rolled out; and
   - QMS processes continually evolve; consequently, associated QMS documentation is revised, which results in a need for retraining.

   QMS training is not necessarily formal; it may in fact be as informal as self-study by employees of applicable QMS documentation.
2. **Monitoring process adherence in real time** — When an organization's quality assurance department personnel participate in, observe, or assess process execution in real time, it provides them with valuable information on the extent of QMS deployment in different parts of the organization. The mere presence of quality assurance personnel during process execution, either as participants or as observers, usually serves as an excellent mechanism for forcing process adherence. Information gained from such firsthand experience can be utilized by the quality assurance department personnel for initiating appropriate corrective action, where needed.
3. **Internal quality audit program** — Finally, a rigorous internal quality audit program wherein all employees are cycled through the quality audit program over a period of time serves as an

excellent way for promoting process awareness and process adherence among employees. Internal quality audits formally close the loop on QMS deployment by helping identify deficiencies in process execution or in defined processes and supporting QMS documentation. These deficiencies then can be addressed by initiating appropriate corrective action.

Although real-time monitoring and internal quality audits might serve as the primary means to verify and promote QMS deployment, there are other means that also provide valuable information to assess the adequacy of QMS deployment. For example, use of measurements, in addition to providing information on the current capability or quality of a process or product, also provides information on adherence to currently defined quality practices. Similarly, project postmortem reviews (explained in Chapter 8), in addition to providing information on deficiencies in organizational processes, also can provide valuable information on failure to execute processes in accordance with requirements.

7.1 EMPLOYEE TRAINING*

In any organization, employees can be expected to execute their tasks in accordance with the defined QMS only if they have the necessary competence for the job, including appropriate training on the applicable QMS processes and supporting documentation. Before we examine employee training, let us examine the broader concept of employee competency.

Competency is the demonstrated ability to apply knowledge and skills, and is derived from education, training, skills, and experience. Skills may include both technical and soft skills. Technical skills pertain to technical skill requirements to perform a job, while soft skills pertain to behavior, attitude, communication, and leadership requirements to perform the job.

Competency needs for different employees in the same role may be different, because different employees possess different competencies. Consequently, the gap (competency needs) between what they need to know to perform their job (competency requirements) and what they already know will vary. To determine the competency needs of an employee, the competency requirements for the employee's job must be known. Competency requirements need to be determined not only to identify competency needs for employees, but also for use as criteria in hiring of new employees, and to determine eligibility for promotion of current employees. The following subsections describe the essential elements of an employee training program.

* Parts of this section are adapted from Vivek (Vic) Nanda, *ISO 9000:2000— Achieving Compliance and Continuous Improvement in Software Development Companies*, ASQ Quality Press, Milwaukee, 2003. With permission.

7.1.1 Identify Employee Competency Requirements

The determination of competency requirements for a position should be made by the supervisor for that position. Competency requirements need not necessarily be identified for each position (or role) in the organization; instead, competency requirements can be identified for a family of similar positions, such as department managers and directors. It is recommended that competency requirements be identified in an employee job description that lists job responsibilities along with required and desired competencies. Desired competencies are "nice to have," but not critical to the job. They may be used during hiring to distinguish between two otherwise equally capable candidates. After the employee is hired, for ongoing professional development of the employee, the employee may be provided training related to desired competencies. If an employee lacks desired competencies, it does not render the employee unqualified for the job. However, if the employee is lacking required competencies, an employee is considered unqualified for the job, and the deficiency must be rectified immediately by providing the employee with the required skills.

7.1.2 Identify Employee Training Needs

Once competency requirements have been identified, the determination of employee competency needs may begin.

There are two sources for the identification of employee competency needs:

1. Competency needs identified at the time of hiring of a new employee, or promotion/transfer of an employee into a new position — At the time of hire, or at the time of promotion or transfer of an employee, the employee and the employee's supervisor should determine what the employee's current competencies are. The employee's current competencies then can be compared with the job competency requirements to determine the employee's competency needs (if any).
2. Competency needs identified during the course of everyday process execution — Some examples:
   - Deficiencies in employee competencies may surface during quality audits.
   - A change in a process may result in the need for retraining of employees.
   - A breakdown in process execution attributed to human failure may result in the need for retraining of employees.
   - Annual performance appraisals of employees typically lead to the identification of new training needs based on employees' planned responsibilities and assignments until the next appraisal.

- An employee might be asked to work on an assignment that requires a particular competency.
- An employee's job responsibilities might change (in which case the employee's job description must also be enhanced to list the new required competencies).
- A new competency might be required for use of a new technique or tool.
- An employee might be in the process of being groomed for another position that requires the new competency.

Once competency needs have been identified, they should be documented in an employee competency gap matrix as shown in Table 16. New competency needs should be added to the matrix as they are identified, and should be marked as completed once addressed.

7.1.3 Take Appropriate Action to Address Competency Needs

Competency needs must be addressed, typically through employee training. However, employee training is not always the best solution to address a competency need, although it is often adopted by organizations as the first solution. All viable options should be investigated, instead of assuming training will rectify all competency-related deficiencies. These options include:

- Outsourcing of certain processes (or tasks) that are beyond the current and planned competency of an organization's employees;
- Modification of processes; and
- Modification of QMS documentation.

When employee training is identified as the most appropriate solution to a competency need, employees may be provided training within the organization or from a qualified outside vendor. Training may be provided by various methods, such as:

- Instructor-led training (ILT) courses
- Web-based training (WBT) courses
- Computer-based training (CBT) courses
- Academic courses provided in colleges as part of a continuing education program (or an advanced degree)
- Self-study of relevant QMS documentation
- Formal mentoring of an employee by a qualified coworker or supervisor
- On-the-job training (OTJ); that is, experience and information gained firsthand from executing assigned tasks

Table 16 Employee Competency Gap Matrix

Department: Receiving Inspection

Employee Name: Elle Spitzer   Employee Title: Quality Inspector   Level: Management ☐ Individual Contributor ☑

#	*Training Name*	*Training Method*	*Vendor*[a]	*Due Date*	*Completion Date*	*Notes*[b]
Technical Skills						
1	Formal reviews and inspections procedure	Self-study	Internal	6/30/04	6/29/04	
2	Quality management system overview	Instructor-Led Training	Internal	~~10/15/04~~ 12/17/04		Due date postponed because QA department postponed October training session to December.
Soft Skills						
	None					
Certifications						
3	IPC-A-610 Worker Proficiency Course	Instructor-Led Training	Electronic Technology Center	7/1/04		

[a] Internal or External (specify company name).

[b] Provide brief explanation if the training is withdrawn for any reason, or if the training is not completed by the due date and is rescheduled, or for other situations requiring explanation.

Although the implementation of an effective in-house training program is a topic of considerable breadth and is beyond the scope of this book, this book includes several examples of documentation required to implement an effective training program. (Refer to Appendix F.) It includes an example of a training course catalog, sample course description, an example of training tracks showing the applicability of different QMS training courses to different areas of an organization, and an example of training material.

7.1.4 Evaluate Effectiveness of Actions Taken

Once employee competency needs have been addressed, effectiveness of the actions taken should be assessed to help identify areas of improvement. Effectiveness may be defined as the extent to which planned objectives are met. There are different ways to determine the effectiveness of the actions taken:

Student testing — A commonly used technique to assess training effectiveness; entails administering the same test to the students before and after training, and comparing the scores to determine the extent to which the students have learned from the training. The assessment of training effectiveness should not be confused with the training evaluation. Although the two overlap in certain areas, they have distinct objectives. Training evaluation involves collection of student satisfaction feedback on the course content, delivery method, instructor evaluation, and training facility. Assessment of training effectiveness entails testing of the students to determine how well the students have learned the training content material.

Quality audits — By definition, quality audits are performed to determine compliance with applicable quality requirements, and to determine adequacy and effectiveness of the QMS. They serve as an excellent means to assess the effectiveness of actions taken to satisfy employee competency needs. Because quality audits entail auditee interviews, examination of documents and records, and observation of activities, they provide ample opportunities for the auditor to identify deficiencies in employee competence.

Measurements program — Use of measurements can determine the effectiveness of actions taken. Use of appropriate measures before and after the actions are taken can help provide evidence of improvement in employee competence. If employees attended a training course to train them on "peer review detected techniques," one can reasonably expect to see an overall increase in the number and severity of defects detected in future peer reviews.

Employee satisfaction survey — An employee satisfaction survey provides valuable information regarding the employee's personal satisfaction with the training provided to them, or the effectiveness of actions taken pertaining to their area of work. For example, if QMS documentation pertaining to customer support employees was created or significantly revised to address employee competence needs, yet customer support employees indicate in the employee satisfaction survey that QMS documentation is lacking or inadequate for them to effectively perform their job, it indicates an area for further corrective action.

7.1.5 Maintain Training Records

In cases where competency needs are addressed by means of employee training, associated training records should be maintained. It is recommended that a separate competency records file be maintained for each employee. This file should be considered non-confidential; it must be open to examination by internal and external quality auditors. Therefore, it should be maintained separately from confidential employee information, such as salary and benefits information. Documents that should be available in the employee's competency file include:

- Job description for the employee
- Competency gap matrix of the employee
- Originals or copies of employee training certificates
- Originals or copies of professional certifications and other pertinent training records

When in-house training instructors perform training, records of the competence of the training instructors should also be maintained.

7.2 MONITORING PROCESS ADHERENCE IN REAL TIME

Once employees have been provided the required competency to execute their tasks in accordance with the defined QMS, the process of monitoring adherence to processes in real time can begin. Real-time monitoring of adherence to the defined QMS entails participation, direct observation, or assessment by the quality assurance department of process execution. This provides them with firsthand information on the extent of compliance to the defined QMS, difficulties encountered with the QMS, QMS limitations, and effectiveness of the QMS, and it facilitates timely corrective and/or preventive action. Examples of means that may be used to monitor the QMS in real time include but are not limited to:

- Audits of processes or projects during execution, or audits of products under development. (Internal quality audits are discussed in the next section.)
- Participation of Quality Assurance (QA) personnel with specific roles during process execution — for example, as a reviewer in formal and informal reviews of artifacts
- Participation of QA personnel in project review meetings where issues pertaining to the QMS often arise (explicitly or implicitly). (Refer to Section 5.2.1.2 for more explanation.)
- Direct observation of process execution by QA personnel acting as silent observers

7.3 INTERNAL QUALITY AUDIT PROGRAM

A quality audit is a systematic and independent examination to determine whether or not activities have been performed in accordance with:

- The requirements of the applicable industry QMS standard; and
- The organization's internal quality requirements (specified in the QMS documentation).

A quality system audit entails an examination of the activities performed and records maintained to determine if the QMS implementation is effective and adequate (to meet applicable requirements). For example, an industry QMS standard may require that an organization perform activity 'x.' In its QMS procedures, the organization may require that in addition to activity 'x,' activities 'y' and 'z' also be performed. Therefore, the quality system audit must verify if all these requirements are satisfied, and the audit report should address:

- Observed nonconformances — that is, situations requiring corrective action. Examples of nonconformances include:
  - Failure to comply with stated process requirements during process execution
  - Failure of the product to comply with product requirements
  - Failure of the organization to meet reasonable customer expectations (which result in a legitimate customer complaint)
- Observed discrepancy or errors in any type of documentation: process, project, or product documentation
- Potential problems (situations requiring preventive action)
- Opportunities for improvement
- Effectiveness and adequacy of agreed-upon corrective and preventive actions implemented in response to earlier quality audits

In auditing terminology, the organization that requests the audit is referred to as the client, the personnel who are audited are referred to as the auditees, and the person who performs the audit is called the auditor. The auditor who leads the entire audit team through the audits and is responsible for all phases of the audit is called the lead auditor (or lead assessor). The auditor must make an objective assessment regarding conformance to the applicable requirements by using such techniques as auditee interviews, examination of records, and observation of activities. The auditor must be authorized by a relevant authority to conduct the audit and qualified (with appropriate formal training) to perform quality audits. There are numerous quality training organizations, such as the ASQ, that offer five-day lead auditor training courses, or two-day internal auditor training courses. It is also required that the auditor be independent of the department being audited; the auditor must not have any responsibility in the areas being audited. This is necessary to ensure that the audit is free of bias and influences that could affect the objectivity of the audit. It is permissible for a department to informally audit itself in order to informally assess its preparedness for a formal audit.

7.3.1 Types of Quality Audits

Quality audits primarily are of two types, internal and external. In addition to quality system audits, there are other types of audits, discussion on which is beyond the scope of this book; however, for the sake of completeness, a brief description follows:

Process audit — This is an audit of a process against its documented description or against the responsible management person's description of the process (if the process is undocumented).

Product audit — This is an audit of a product against applicable specifications.

Project audit — This is an audit of a project against applicable project requirements and project plans.

Let us examine internal and external quality audits in more detail:

Internal quality audit — Internal quality audits are referred to as first-party audits because qualified employees from the organization conduct these audits. These employees typically belong to the quality assurance department and are qualified in terms of:

- Appropriate quality auditor training (and certification, when desirable); and
- Adequate experience and expertise in the areas audited.

A quality auditor must be independent of the area audited. For this reason, the quality assurance department typically reports directly to the person with overall management responsibility of the organization, such as the company's president. This helps ensure the impartiality of the quality assurance department and protects it from bias and influence that otherwise could compromise the objectivity of the auditors.

Some small organizations that do not have qualified auditors in-house engage the services of a quality consulting company to perform periodic internal quality audits. This practice is much less desirable, because if the organization truly were committed to quality, it would be a serious contradiction and handicap if it were not to have qualified quality personnel in-house to assess whether or not its processes meet quality requirements, and if corrective or preventive action is required. Such expertise is fundamental for sound auditing skills; therefore, generally speaking, an organization cannot afford not to have qualified internal quality auditors on its staff.

External quality audit — External quality audits are either second- or third-party audits. A second-party audit is a supplier audit; i.e., an audit of an organization by its customer, or an audit by an organization of its supplier. A third-party audit is an audit of an organization by an independent third party, such as a QMS registrar. Depending on the criticality of each supplier in terms of its impact on the quality of the product or service delivered by the organization, it is strongly recommended that the organization conduct second-party audits of its suppliers. Such audits enable an organization to exercise effective control over its suppliers' processes. Supplier audits should be supplemented by other appropriate means to control supplier performance, such as periodic meetings with the supplier to review supplier performance and discuss quality issues, and joint review of established quality metrics to assess and improve supplier performance.

The overall auditing process for the different types of aforementioned audits is essentially the same and is described next. There are some implementation-level differences between these different types of audits, discussion on which is beyond the scope of this book.

7.3.2 Audit Scheduling

At the beginning of the year, the quality management representative should ensure that an annual internal quality audit schedule is prepared that lists the internal quality audits planned for each month. Such an audit schedule can be a simple spreadsheet with separate columns for each month and separate rows for each planned audit. The audit schedule should be provided to management personnel with responsibility for each area to be audited.

Besides planning for internal quality audits to be conducted by the quality assurance department, the organization should plan for an audit of its internal quality audit program and of the quality assurance department. Such an audit must be performed by personnel other than those responsible for the internal quality audit program. Some organizations engage the services of outside quality consultants to perform such an audit.

In addition to planned internal quality audits, the QMS should include provisions to allow unannounced audits, if necessary. The benefit of not preannouncing an audit is that it promotes a culture of employee preparedness for internal audits at all times (by continuously adhering to applicable requirements), as opposed to employees scrambling to get prepared for an announced audit in an effort to forestall a potential audit finding by temporarily rectifying nonconformant practices (for example, discarding uncontrolled QMS documentation). Unannounced internal quality audits pose scheduling challenges related to possible interference with critical activities that may be under way and possible conflicts with availability of required auditees, and human challenges arising from the possibility of promoting a feeling of fear and apprehensiveness towards the internal quality audit program.

Considerations and criteria that should be used in establishing the annual internal quality audit schedule include:

- The requirement that each major element of the QMS (or each separate clause in the applicable industry QMS standard) be audited at least once a year; when the QMS is newly established, more frequent audits typically are required.
- The requirement that critical areas of the product development process and areas of weakness (as determined from past audits) be audited at least twice a year
- Consideration should be given to not scheduling internal audits during peak holiday and vacation times of the year.
- Consideration should be given to scheduling an audit of areas that have undergone recent change (for example, changes in operating processes and procedures, organization changes, and changes in departmental responsibilities).
- The nature and severity of past audit findings in each audit area should be considered in planning future audits of those areas.
- For organizations that are (or intend to be) formally registered to an industry QMS standard, consideration should be given to not scheduling internal audits at times that conflict with the audit schedule of the registrar.

7.3.3 Audit Planning

As the due date for an audit draws closer, the quality management representative should identify a suitable lead auditor who has overall responsibility for planning, executing, and reporting of the audit. The lead auditor should begin planning for an audit by examining the audit scope to determine if additional auditors are required. The audit team reports directly to and executes tasks assigned by the lead auditor. It is the responsibility of the lead auditor to ensure that the auditors are suitably qualified, with appropriate training, experience, and subject matter expertise relevant to the audit. Occasionally, the lead auditor may also select observers (auditors-in-training) who participate as silent observers in the audit team.

The lead auditor is responsible for providing reasonable advance notice about a planned audit to the responsible management personnel by releasing a formal audit plan. It is generally recommended that the audit plan be released at least a month before the planned audit date. The audit plan should contain such information as:

- Objectives and scope of the audit
- Areas to be audited
- Date of the audit
- Lead auditor (and other audit team members)
- Method of conducting the audit

Management personnel should be requested to confirm that the proposed audit dates are acceptable and that there are no special circumstances that necessitate rescheduling of the audit. If an audit involves travel to another site, travel arrangements should be initiated at this time. It is also worthwhile to emphasize in this notification that the purpose of the audit is to audit the QMS, processes, products, or projects (as the case may be), and that the audit must not be viewed as personnel assessment.

7.3.4 Audit Preparation

Once an audit plan has been released, the lead auditor should oversee the preparation for the audit by ensuring that the auditors prepare by reviewing the following documentation:

- Relevant QMS requirements, process documentation, and project documentation
- Relevant past audit reports and internal corrective and preventive action records to ensure past problems are not recurring, and to

identify outstanding problems that need to be followed up during the audit
- Measurement results and trends to identify issues requiring investigation during the audit
- Customer-specific contractual obligations, to verify that they are being met

The auditors should include all items selected for investigation in a formal audit checklist.* Audit checklists serve as an invaluable tool to structure the audit interviews and ensure adequate coverage of all identified questions. An example of an audit checklist is shown in Appendix E.

At least two weeks before the audit, an audit notification should be sent to the respective department management personnel, providing information on:

- The date and time that the audit will be conducted
- The purpose and scope of the audit
- Specific employees requested for the audit (if appropriate)
- The reserved meeting rooms for the audit (if any)
- Any other relevant information

The audit notification should be accompanied by audit checklists for distribution to the department personnel. These audit checklists may be standardized audit checklists that do not necessarily contain the list of specific items selected for investigation during the audit. Distribution of audit checklists helps promote acceptance of audits as a means for free and open sharing of information with the auditor for continuous improvement, as opposed to promoting a feeling of apprehension and discomfort by auditing employees with secret audit checklists kept close to the auditor's chest. Distribution of audit checklists helps make internal audits nonthreatening to the employees and helps put them at ease during the audit interviews.

7.3.5 Opening Meeting

When appropriate, an internal audit should commence with a brief opening meeting with management personnel from the areas to be audited.

* Generally, audit checklists are standardized, and contain questions derived from requirements applicable to the area to be audited. For a particular audit, they can be augmented with additional questions related to items specifically selected for investigation during the audit. In essence, such additional questions are included on an as-needed basis, and they may not result in a change to the standardized audit checklist.

(If appropriate, auditees may be invited to the opening meeting.) The primary purpose of the opening meeting is for the lead auditor to brief the audience about the audit, and to provide them an opportunity to ask questions. At the opening meeting, the lead auditor should:

- Clarify the purpose and scope of the audit
- Provide information on how the audit will be conducted
- Communicate audit times and breaks, and meeting rooms for the audit interviews
- Provide information on time and place for the closing meeting, where the audit results will be presented
- Invite meeting attendees to ask any questions that they might have

When an opening meeting is held, a list of meeting participants should be maintained, along with notes from the meeting.

7.3.6 Audit Interviews

Audit interviews can begin immediately after the completion of the opening meeting. There are two widely used approaches for conducting audit interviews:

Trace forward — In this approach, the auditor audits activities and process inputs at the beginning of a process and then sequentially traces forward the process inputs and activities to the next activity and outputs in the process. In this manner, the auditor audits a process by incrementally stepping forward in a process until the last activity (and final outputs) of the process.

Trace backward — In this approach, the auditing process is reversed. The auditor first audits the final process outputs and then systematically traces them back to the preceding activity and process inputs that produced the outputs. In this manner, the auditor audits a process by incrementally stepping backward in a process all the way back to the first activity (and inputs) in the process.

There are three generally accepted methods for conducting an audit:

- Audit interviews;
- Observation of activities; and
- Examination of objective evidence, such as process outputs and supporting documentation. (When sampling objective evidence, a sample size of three to five is typically recommended.)

In conducting the audit, the internal auditor should bear in mind that conducting internal audits can be and usually is a difficult task because the internal auditors are auditing fellow workers. Some resistance and defensive posturing are natural and expected when people are faced with a situation wherein their work methods and output are evaluated. The auditor should not in any way intimidate the auditee or behave in a way that would reinforce a negative attitude towards the audit. Audit questions should be open-ended, context-specific, or investigatory, as opposed to closed-ended or tricky. The auditor should keep adequate notes from the audits, including the names of the persons audited.

Once a nonconformance is identified during an audit, the auditor should share it with the auditee and ensure that the auditee understands the nonconformance. Identification of the severity of a noted nonconformance may be performed outside of the audit interview because it might require further examination of evidence and consultation with other audit team members. Typically, organizations classify noted nonconformances as major or minor.

A major nonconformance may arise due to:

- A serious deficiency that would adversely affect the quality of the product or service;
- A significant nonconformance against applicable requirements that indicates a complete absence or breakdown of compliance to the requirements; or
- Evidence of system-wide deficiencies.

A minor nonconformance is an observed temporary omission, e.g., an isolated case of nonconformance with a specific requirement, or a minor discrepancy from a stated requirement. Because evidence of system-wide deficiencies constitutes a major finding, several minor findings in a certain organizational process can have a serious cumulative effect, and therefore can be grouped together under a single major finding.

Organizations generally use a third classification, called an observation, to identify a potential problem or an opportunity for improvement. In all cases, nonconformances and observations should be supported with objective evidence. Additionally, the auditor may record positive observations to recognize observed best practices that management personnel may evaluate for appropriateness for use in other parts of the organization.

7.3.7 Closing Meeting

The audit interviews conclude with a closing meeting with the management personnel from the areas audited. (If appropriate, auditees may be invited to the closing meeting.) The primary purpose of the closing meeting is to debrief the audience about the audit, any difficulties or issues encountered during the audit, and results of the audit, and to provide information on the overall adequacy and effectiveness of the processes audited. It is preferable to provide formally documented nonconformances and observations at the closing meeting. (Refer to the Corrective Action Request Form in Appendix D.2.) The attendees of the closing meeting should be informed when the audit report will be formally issued, and of the planned next steps for corrective action planning and implementation. As with the opening meeting, a list of meeting attendees should be maintained along with notes from the meeting.

7.3.8 Audit Reporting

Within approximately two to four weeks of an audit, the lead auditor should issue the audit report to the management personnel of the areas audited. The purpose of the audit report is to describe the audit and its results accurately. Typical contents of an audit report include the following:

- Brief description of the purpose and scope of the audit
- Pertinent details from the audit plan
- Statement of adequacy and effectiveness of the areas audited
- Audit outcome (including a summary of audit findings)
- List of attendees at the opening meeting and summary of what was discussed
- List of attendees at the closing meeting and summary of what was discussed
- Appendices with formally documented nonconformances and observations

The audit report should also describe nonconformances discovered during the audit that were resolved during the audit with a satisfactory corrective action. The audit report should be provided to the management personnel of the areas audited, and shared with the auditees. The management personnel should be provided a due date for a formal corrective action plan. The lead auditor should review the corrective action plan for acceptance prior to authorizing its implementation.

7.3.9 Audit Follow-Up and Closure

Once the auditees have implemented the agreed corrective and/or preventive actions within the agreed time frame, the auditor must verify that the implemented actions adequately and effectively address the identified root causes of the problem. Nonconformances and observations may be closed upon examination of new objective evidence that demonstrates that the problem has been resolved, or by conducting a reaudit of the affected areas. The audit report should be formally closed after all previously noted nonconformances and observations have been resolved.

At all times, the quality assurance department should maintain up-to-date information on the status of each audit performed, and monitor that all audit-related activities are being completed within required time frames. The QMS should specify the retention time and location of all audit records, such as annual audit schedule, audit plans, audit checklists, and audit reports.

PHASE V

IMPROVE

8

CONTINUAL IMPROVEMENT

The continual improvement phase is the final phase of QMS implementation. It entails the use of appropriate mechanisms for continual improvement of the QMS. As explained in Chapter 2, some of these mechanisms might have been defined and deployed in the previous phases, while some might be defined and deployed in the continual improvement phase. This chapter presents various mechanisms an organization can use for continual improvement. It introduces the concept of the continual improvement cycle for realizing continual improvement over successive projects. The chapter concludes with a discussion on how to make continual improvement stick — how to deploy continual improvement gains such that the improved processes constitute the new baseline (or standard) for process execution.

8.1 ACHIEVING THE GAINS: MECHANISMS FOR CONTINUAL IMPROVEMENT

There are various mechanisms that may be used for continual improvement in an organization. None of these mechanisms is sufficient by itself as a means for continual improvement. Use of at least some if not all of these mechanisms is necessary to have sufficient tools for continual improvement.

8.1.1 Quantitative Process Improvement*

Quantitative process improvement involves the use of process measurements to analyze process performance, to define quantitative goals for

* This section is adapted with permission from Nanda, Vivek (Vic), *ISO 9001 — Achieving Compliance and Continuous Improvement in Software Development Companies*, ASQ Quality Press, 2003.

process improvement, and to identify actions necessary to achieve the defined goals. Process measurement helps quantify the current capability of a process and the desired capability of a process. Awareness of both of these is essential for continual process improvement.

Institutionalization of quantitative process improvement in an organization requires the establishment of a measurements infrastructure. Items that should be addressed during establishment of a measurements infrastructure include:

Measurement item — The measurement item is the entity being evaluated by means of measurements to identify opportunities for improvement. The measurement item may be the organization itself, or processes, projects, and products. Project measurements are required because in the context of product development, processes typically are executed not in a vacuum, but within the bounds of a product development project. Project measures complement process measures and are useful during project planning, for real-time project tracking and control, and for conducting project postmortem analysis to identify future process improvements. Process measurements are used for monitoring and improving QMS processes, and product measurements are used for monitoring and improving product quality. Sometimes the same metric may belong to more than one metric type: process, project, or product metric.

Goal-question-metric approach to metrics definition — The goal-question-metric (GQM) approach to identifying metrics is based on the premise that for metrics to be purposeful, they should be derived in a top-down fashion [Bas84]. This entails a three-step process:

1. A measurement goal or purpose should be determined.
2. A set of questions should be asked to further define, or characterize, the measurement goal.
3. Metrics should be defined to support each question.

Sometimes the measurement goal may be derived by decomposition of a higher-level goal, as in the case of management by objectives techniques where high-level organizational goals are decomposed at lower levels of the organization. Also, bear in mind that an organization collects measurements for use by different users, such as senior management, project managers, process owners, engineers, quality department personnel, and customers. In determining what characteristics to measure and what metrics to use, consideration should be given to addressing the needs of various types of users who would be interested in a specific measurement.

The following example illustrates the GQM approach to metrics definition:

An organization identifies the following high-level goal (G): Improve project planning. A valid set of questions to help characterize project planning is:

1. What is the current accuracy of estimating a project schedule?
2. What is the current accuracy of estimating project effort?
3. How does the actual cost of project execution compare with the projected cost?

For each of these questions, the following corresponding project metrics can be identified (refer to the next subsection, "Measurement Logistics," for further guidance on metrics definition):

M1. Planning Precision (PP): 100–100 * | (Planned lead time – Actual lead time)/Planned lead time | =%

Note: If PP is negative, it is set to zero (0). This happens when Actual lead time > 2 * Planned lead time.

M2. Effort Estimation (EE): 100–100 * | (Planned effort – Actual effort)/Planned effort | =%

Note: The measurement unit for effort is person-hours. If EE is negative, it is set to zero (0). This happens when Actual effort > 2 * Planned effort.

M3. Cost Performance Index (CPI): Actual cost of work performed/Budgeted cost of work performed

Note: CPI = 1 (project is within budget); CPI > 1 (project is over budget); CPI < 1 (project is under budget).

Measurement logistics:

1. **Metrics definition** — It is recommended that each of the identified metrics possess certain desirable characteristics. The identified metrics should be:
   - Useful
   - Widely accepted as a good indicator of the characteristic being measured
   - Precisely defined
   - Simple to understand
   - Relatively easy to compute
2. **Primitive data collection method and tools** — Most metrics are derived metrics, as opposed to primitive metrics. The final

measurement value is computed (derived) from certain primitive data. For example, computation of planning precision requires that primitive data for the planned lead time and actual lead time be collected. Therefore, for each identified metric, the method and tools for collecting required primitive data should be specified. The measurement unit for the primitive data also should be defined and consistently applied.

3. **Rules for computing measurements** — The computation of a measurement value, hereafter referred to as measurement data, should be free from personal interpretation and subjectivity. Measurement data that has been compromised, either intentionally or unintentionally (due to lack of defined rules for computing it), will render it inaccurate, worthless, and open to discussion. Therefore, all dos and don'ts associated with computing measurement data should be established and agreed to by the appropriate parties.
4. **Measurement data storage** — Measurement data should be stored in a centralized measurements repository or database, or a collection of databases. The storage location for measurement data should be established and communicated to appropriate parties.
5. **Measurement data reporting** — Guidelines should be established regarding:
   - How often measurement data are to be reported;
   - Who reports measurement data and to whom; and
   - How measurement data are to be reported.

   Reported measurement data generally should be supported by trend and causal analysis of the data. When the reported data deviate from the established upper and lower control limits (in the case of a control chart), or fall below a predetermined target value, remedial steps should be taken by means of appropriate corrective, preventive, and improvement actions.
6. **Roles and responsibilities** — Roles and responsibilities should be identified for all activities within the measurements infrastructure, beginning with identification of and agreement on required metrics, and ending with analysis of measurement data for future improvements. If the responsibility to collect primitive data is distributed among various departments in the organization, the responsibility for ensuring cross-functional consistency in measurement practices should be assigned to one department. Typically, the organization's quality assurance department fills this role.
7. **Training** — Training and ongoing consulting support should be provided to all personnel involved in identifying needed metrics, collecting primitive data, computing measurements, reporting, and

analyzing measurement data. This includes training on the use of quality tools for data analysis, which are described next.

Analyze measurement data for future improvements —

1. **Analyze current measurement data** — Once measurements have been computed, they must be analyzed, both for trends and for reasons behind a particular trend or deviation from control limits or target values. Examples of quality tools that might be used for data representation and analysis include control charts, run charts, scatter diagrams, Pareto charts, ishikawa (or fish-bone) diagrams, tree diagrams, and affinity diagrams [Fos00].
2. **Determine SMART target values** — The basic premise of the GQM approach to metrics definition is that in order to define useful metrics, there has to be a measurement goal. It follows that once a metric has been defined, in order to improve performance from current levels, or to maintain performance within certain bounds, the original qualitative measurement goal has to be quantified in terms of the defined metric. To maintain performance within certain bounds, organizations typically use control charts with specific upper and lower control limits derived from past data — a mechanism known as statistical process control. However, when process improvements are to be achieved, a new target value needs to be identified. The new target value should possess the SMART characteristics described earlier in this book.

   For example, the original qualitative measurement goal of "G: Improve project planning" discussed earlier in this section should now be supported by underlying quantitative goals such as:
   a. Improve planning precision from the current value of 70% to 80% by the end of the next calendar year;
   b. Improve effort estimation from the current value of 60% to 75% by the end of the next calendar year; and
   c. Improve cost performance index from the current value of 1.25 to 1.15 by the end of the next calendar year.

   In order to ensure that an established target value possesses SMART characteristics, the target value should be identified by involving appropriate personnel from relevant areas, as opposed to arbitrarily determining the target value and foisting it upon them. For an internal process, a quantitative improvement goal should be established as per internal customer needs, and after careful consideration of the current capability of the process. For a customer interfacing process, a quantitative improvement goal should be derived from the relevant organizational goals — this is the concept of deployment of corporate quality objectives (explained in the next section).

The identified target values should be deployed within the organization by tying them to individual and team annual performance objectives, when appropriate. Employees should be provided individual and team-based reward incentives to achieve the identified target value for each metric. Care should be taken to ensure that the identified goals are balanced goals that do not promote the achievement of a specific objective at the expense of quality and other considerations. For example, a measurement goal of 90% on-time product delivery will invariably be at the expense of quality considerations unless there are complementary quality goals that must be met, and there are built-in accountability mechanisms in the reward system to preclude the organization from persistently delivering an inferior quality product on time. In such an environment, the quality goals for a product release have a semblance of having been met, but customer-reported defects and experiences with the product contradict organizational claims about the true quality of the product.

3. **Identify improvement actions** — In order to facilitate the achievement of the new target value, specific improvement actions must be identified and assigned to responsible personnel. Without the identification of improvement actions, realization of the target value is unlikely to materialize, because the current process will only result in achievement of current performance levels. Similarly, when performance deviates from established control limits, appropriate corrective actions should be identified, by means such as ishikawa diagrams (for identifying the root cause) and tree diagrams (for identifying the means to address the identified root cause).

8.1.2 Corporate Quality Objectives

Long- and short-term quality objectives define the strategic and tactical objectives for quality improvement in an organization. The implementation of suitable improvement actions necessary to achieve these objectives serves as an indispensable means for continual improvement. The challenge most organizations face pertains to the implementation of the identified objectives: How should the high-level quality objectives be systematically decomposed into lower-level quality objectives (and related improvement actions), so that the achievement of the lower-level quality objectives contributes to the achievement of the higher-level quality objectives? The solution lies in a systematic approach called policy deployment, or strategy deployment.

Policy deployment is the process of deploying the goals and initiatives of corporate top-level management down to the lowest levels of the corporate hierarchy. The purpose of policy deployment is to manage and achieve significant improvements in a company's performance. Policy deployment enables an organization to involve its employees in planning how their work contributes towards achievement of the organizational goals, increasing their commitment to achieving those goals, which increases the likelihood that the goals (and the underlying strategies reflected in these goals) will be achieved. An example of a process for deployment of corporate quality objectives is provided in Appendix C.7.

8.1.3 Internal Quality Audits

Internal quality audits are a valuable tool for continual improvement because their primary purpose is to determine the adequacy and effectiveness of the QMS, and to report situations that require corrective or preventive action or that otherwise present an opportunity for improvement. (Refer to the explanation of internal quality audits in Chapter 7.)

8.1.4 Corrective and Preventive Action Requests

Strictly speaking, a corrective or preventive action does not always qualify as an example of continual improvement. For instance, by definition, a corrective action is required when there is a:

Failure to conform to process or product requirements — Process execution does not conform to the process definition (which may not be documented), or the product does not conform to product requirements. The nonconformance in this case arguably is avoidable. Reasons for such nonconformances include personnel negligence, resource constraints, schedule pressure, or other factors (most of which can generally be avoided or mitigated).

Deficiency in the defined process — The nonconformance arguably is unavoidable because the defined process is inherently deficient.

In case of a failure in process execution, corrective action planning entails a root cause analysis of why the process was not executed in accordance with its definition, and what actions should be taken to correct the observed situation and to enforce process adherence in the future. In essence, the corrective action plan seeks to address the nonconformance and restore the process to its planned capability (or performance level).

In the case of a deficiency in a defined process, corrective action planning entails an examination of the deficiency that needs to be resolved. In essence, the corrective action plan seeks to enhance the capability or performance level of the process, resulting in improvement from the earlier capability of the process.

A preventive action might not always qualify as an example of continual improvement if it does not result in an improvement in the capability of the process. Nevertheless, it is obvious from the foregoing explanation that corrective and preventive actions warrant discussion as a means of continual improvement.

A situation requiring corrective or preventive action might be detected during an internal or external audit, or during the normal course of business. In both cases, most companies employ a similar process for implementing corrective and preventive actions* (refer to Appendix C.5):

The observed situation is recorded by the auditor, typically a member of the quality assurance department, in a corrective action request form (refer to Appendix D.2) or a preventive action request form, as appropriate. This form is provided to the appropriate management person with a request for a corrective (or preventive) action plan. Note that an ordinary employee may identify a situation warranting corrective or preventive action, and may then communicate it to the quality assurance department for further action.

The responsible management person from the affected area responds with a root-cause analysis** and proposed corrective or preventive action plan. The auditor reviews the proposed corrective or preventive action plan to determine if it effectively addresses the identified root causes of the problem. A corrective action plan should not only describe how the noted problem will be corrected, but, if appropriate, how it will be prevented in the future***. At this time, both parties negotiate a time frame within which the actions must be implemented.

In the case of an observed nonconformance, it might be agreed that pursuing a corrective action would yield limited or no gains, and therefore it may be appropriate not to pursue a corrective action. For example, a

* When a nonconformance (or potential nononformance) is observed in a product, it usually is handled by a separate defect management process for the elimination of known product defects or for preventing potential product defects.

** In the case of a situation requiring preventive action, in reporting a potential nonconformance, the auditor might identify the root cause that is likely to result in a potential nonconformance. This is because, by definition, a potential nonconformance has not yet occurred, but an astute auditor or employee may be able to anticipate a potential nonconformance from examining a particular situation. The identified situation itself might be the root cause of a potential nonconformance.

***As an example, if a certain type of defect is found in a product, corrective action planning must address not only the near-term objective of correcting the observed product defect but also the longer-term objective of preventing the introduction of similar defects into the product.

nonconformance might have been found against a process that already is planned to be discontinued or significantly revised, or a nonconformance might have been found long after process execution was completed. In the latter case, correcting the nonconformance would involve significant costs with little or no gains. In such a situation, it may be advisable instead to pursue a preventive action for the problem to preclude its recurrence in the future.

The planned action is then implemented within the agreed time frame, and upon completion, the auditor verifies records of the implemented actions. If the action is implemented as agreed, the corrective or preventive action request is formally closed off.

The aforementioned mechanism, or a similar mechanism, should be utilized for handling customer complaints as well. (Refer to the related discussion on handling of customer complaints in the Customer Support section in Chapter 5.)

8.1.5 Project Postmortem Review

The project postmortem review (also called lessons learned meeting) is conducted upon completion of a product development project to discuss lessons learned from the project, including:

- Opportunities for improvements — problems or deficiencies encountered during a project; and
- Positive experiences — experiences that had an overall positive influence on the outcome of a project.

The purpose of the project postmortem review is:

- To initiate appropriate corrective and preventive actions to address the opportunities for improvement so that the underlying problems or deficiencies are resolved; and
- To initiate appropriate action to ensure that the positive experiences are replicated in future projects (where applicable) — for example, by making needed process and/or procedural changes.

An example of a procedure for conducting and following up on a project postmortem review is provided in Appendix C.6.

8.1.6 Improvement Suggestions from Employees

Any organization serious about continual improvement should solicit suggestions for QMS improvement from employees, as they have firsthand information regarding the capability and limitations of each process, difficulties encountered in process execution, and opportunities for improvement throughout the organization. Moreover, active solicitation of employee input for continual improvement sends a clear message that the organization values employee feedback, and helps encourage employee participation in the maintenance and improvement of the QMS. Employees should be reminded by appropriate means and at appropriate events to submit their improvement suggestions. Suitable incentives should be offered to encourage employees to submit their improvement suggestions, such as recognition and rewards for most outstanding improvement suggestions. Information and statistics on employee improvement suggestions already implemented also should be shared with employees; this helps demonstrate that the organization is serious not only about collecting employee suggestions, but also about implementing the accepted suggestions. An example of a procedure that may be used for handling process improvement suggestions submitted by employees is provided in Appendix C.2.

8.1.7 Customer Satisfaction Surveys*

The premise of using customer satisfaction surveys as a tool for continual improvement is that if an organization's QMS is adequate and effective, it should result in satisfied customers. Conversely, if an organization's customers are dissatisfied, it most likely is a consequence of deficiencies in the organization's QMS. Even in organizations where the overall customer satisfaction rating is deemed "acceptable," opportunities for improvement of the QMS exist. Therefore, an organization needs to complement its internally driven vehicles for continual improvement, such as internal quality audits, employee suggestions, and project postmortem reviews, with externally driven vehicles for continual improvement, such as customer satisfaction surveys, in order to leverage internal and external sources. Measuring customer satisfaction makes good business sense because it is cheaper to retain current customers than to attract new ones, and loyal customers offer high long-term returns to their suppliers and help attract new customers.

While it is necessary to assess customer satisfaction or dissatisfaction for continual improvement of the QMS, customer dissatisfaction might result from reasons beyond the scope of an organization's QMS. Customer dissatisfaction might result from unrealistic expectations established by the

* This section is adapted with permission from Nanda, Vivek (Vic), *ISO 9001 — Achieving Compliance and Continuous Improvement in Software Development Companies*, ASQ Quality Press, 2003.

sales personnel prior to contractual negotiations for a new product, or from the technical direction and future evolution of an organization's product being incompatible with a customer's needs.

In order to systematically collect, analyze, and act upon customer satisfaction data, a customer satisfaction measurement process first should be defined. An organization may also choose to engage the services of a third-party vendor for preparing the survey questionnaire (with organizational input), for conducting the survey, and for reporting survey results, although the organization still would have the responsibility of analyzing the survey data and devising a corrective action plan. The scope of the customer satisfaction measurement process should cover the following major phases:

1. Planning
2. Execution
3. Analysis and corrective action planning
4. Results reporting and corrective action implementation
5. Follow-up

Each of these phases is briefly described:

Planning — This phase entails preparing the customer satisfaction survey questionnaire and planning how to administer the survey to customers. In preparing the questions for the survey, the organization should consider what issues are important from a customer perspective. The survey also should contain questions to elicit customer feedback on possible areas of weakness that are discernible from customer satisfaction data. For example, if the organization is concerned about the user-friendliness of its product, appropriate questions should be devised to elicit customer feedback. The organization also should include questions that help it validate the acceptability of results of its quality metrics, and help in establishment of target values for quality metrics. For example, if the organization uses metrics to measure responsiveness to the customer, it can use the customer satisfaction survey to investigate whether or not the current level of responsiveness is acceptable to the customer, and to identify how much improvement is needed. Preferably, the survey also should elicit customer recommendations for organizational improvement in free text format.

The survey should be constructed so that the questions are categorized in logical groups, such as product quality and quality of customer service. There should be an appropriate response scale for each question. During this phase, decisions also should be made regarding survey administration (by mail, the phone, or the internet), who will administer the survey, how many and which customers to contact, whom to contact at the customer locations, and other such administrative items. Finally, the survey should be pretested to gauge respondent's reactions to the survey layout and content. It should be administered to a few employees not involved in the development of the survey, or it may be pretested with a small set of customers. Upon completion of the pretest phase, revisions should be made to the survey as needed.

Execution — During this phase, the customer satisfaction survey is administered to the preidentified customers as per the identified means. If the customers are required to provide their responses by mail or the internet, a reasonable time frame for accepting the responses should be established. The results of the customer satisfaction survey should be compiled in a formal report with an executive summary. The report should provide a summary of:

- Overall customer satisfaction survey rating
- Customer satisfaction rating by customer (if appropriate)
- Customer satisfaction rating for each question and group of questions, such as product quality

When customer comments on recommendations for organizational improvement are collected, they should be grouped together logically along with an indication of how many customers (or what percentage of respondents) made a similar recommendation. (This should be used for prioritizing corrective actions.) This phase should end with an immediate acknowledgement to the customer satisfaction respondents. Respondents should be thanked for their time completing the survey, and advised on the next steps that the organization plans to take to address the survey findings — including a promise of future communication to inform them of the planned corrective actions. Such an acknowledgement may be sent by a simple postcard, or by a letter signed by the organization's quality management representative.

Analysis and corrective action planning — Once the customer satisfaction survey results have been compiled, an initial presentation may be arranged to share the results of the survey with management personnel. This may help in obtaining management input and support for required resources and personnel for corrective action planning and implementation. Root-cause analysis and corrective action planning should be conducted by including the appropriate departments that are involved in the specific item or process being assessed. The established corrective action plan should be timebound, with clear assignment of responsibility for each action.

Results reporting and corrective action implementation — Once the survey results have been analyzed and a timebound corrective action plan has been formulated, the information should be communicated to management personnel to secure their approval for the planned corrective actions. The survey results and planned corrective actions should be shared with employees. Employee participation is essential in implementing the corrective actions. By being made aware of the survey results and causes of observed problems, employees will be better able to understand the need to change their processes and documentation in accordance with the corrective action plan. A summary of the survey results and associated corrective action plans also should be shared with the customers. This helps positively reinforce that the organization is committed to improving customer satisfaction levels, and is taking specific steps to address the results from the survey.

Corrective action implementation progress must be monitored continually, and implementation progress must be reviewed at periodic management reviews of the QMS. Such status reporting to senior management helps ensure that all departments accord high priority to their respective tasks due to senior management awareness of their progress. Moreover, it enables timely management intervention if the implementation of the corrective action plan starts to fall behind schedule.

Follow-up — This is the final phase in the customer satisfaction process, and it closes the loop on the implemented corrective actions by verifying their effectiveness in achieving planned results. The results from the next customer satisfaction survey should be analyzed to ascertain whether there has been improvement from previously reported satisfaction levels. If corrective actions were implemented for a particular item or process, but there is no improvement in customer satisfaction or customer satisfaction has regressed, additional corrective actions typically would be needed to supplement those already implemented. This might require a reanalysis of the root causes and required remedial action.

8.2 HOW IT ALL TIES TOGETHER: THE CONTINUAL IMPROVEMENT CYCLE

Having described some of the tools for continual improvement, let us examine how these tools fit in the overall context of continual improvement, and how a product development company's processes are improved incrementally over successive projects — a concept known as the continual improvement cycle that is introduced in this section (Figure 25). Understanding of the continual improvement cycle helps one understand how continual improvement results as a natural consequence of the never-ending process of incremental improvements to an organization's processes, use of the enhanced processes for subsequent process execution, and identification of future process changes based on experience from

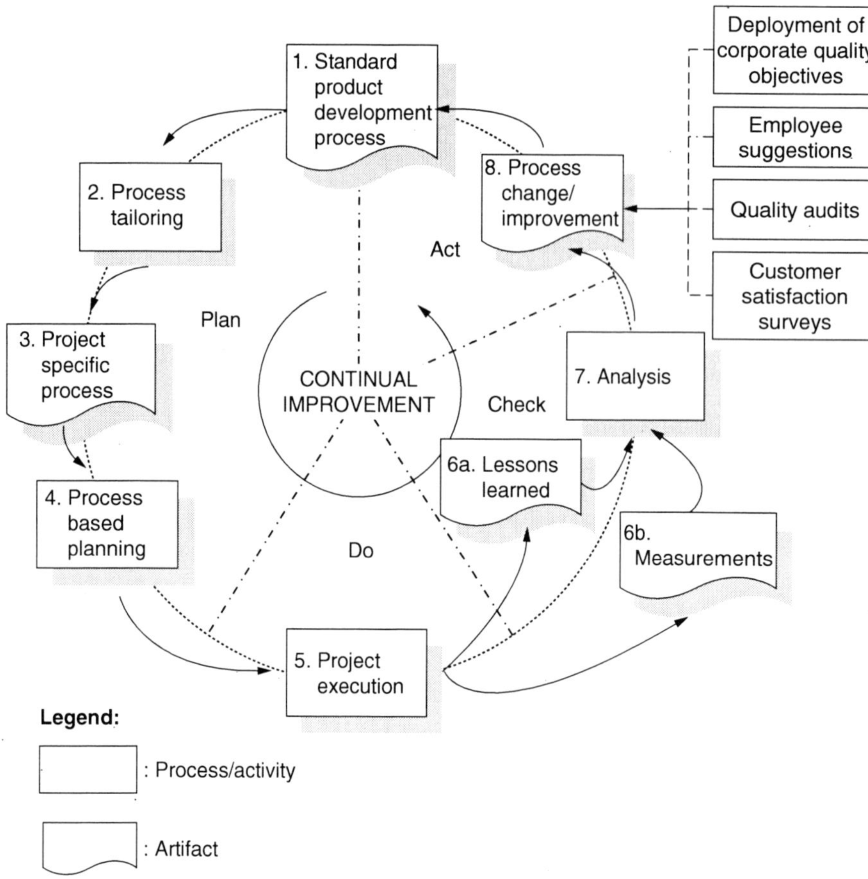

Figure 25 The Continual Improvement Cycle

use of the processes. An understanding of continual improvement, which often is regarded as an abstract concept at worst or an overarching concept at best, enables one to tie this concept to execution of product development projects. This understanding is invaluable for an organization's employees to make the transition from theory to practice — to understand how the concept of continual improvement relates directly to their day-to-day jobs.

The following are the key steps in the continual improvement cycle:

- At the start of a product development project, during quality planning for the project, the standard product development process is tailored for the specific project by identifying the list of planned deviations from the process, along with supporting rationale and risk mitigation. (Refer to the discussion on process tailoring in Chapter 4, and on quality planning in Chapter 5.) The outcome of the process-tailoring exercise is a list of planned process deviations which, taken together with the standard product development process, define the process to be followed for the specific project (project-specific process).

- The project-specific process forms the basis for executing project planning activities such as preparation of Gantt charts, preparation of Program Evaluation and Review Technique (PERT) charts, critical path analysis, effort estimation, and resource allocation (process-based planning).
- Once formulation of project plans based on the project-specific process is complete, project execution can begin. After completion of product development and validation activities, the product is authorized for release to customers if all release criteria have been satisfied. The project then formally concludes.
- A project postmortem review (lessons learned exercise) is organized soon after the conclusion of a project, to identify lessons learned and to initiate corrective and preventive action.
- Project performance also is assessed by analyzing the process, product, and project measurements gathered to identify opportunities for process improvement and to initiate suitable corrective and/or preventive action.
- Processes are changed as needed to address the observed deficiencies uncovered by the project postmortem review and measurement data analysis. Corrective and preventive actions arising from a quality audit, employee suggestion, or customer satisfaction survey also might result in a process change. Similarly, a process also might be changed as part of an improvement action undertaken for the deployment of a corporate quality objective.
- For each process change, a decision is made to either pilot or deploy the process change in the next project. In cases where a pilot of a process change is deemed necessary, results of the pilot are reviewed to determine if the process change should be deployed as is, deployed (or repiloted) with modifications, or withdrawn. Once a decision to deploy a process change is made, relevant QMS documentation is revised in order to formalize the process change.
- The revised product development processes and supporting QMS documentation then form the basis for process tailoring and planning for the next project, and the cycle is repeated.

8.3 SECURING THE GAINS

Once an organization has improved a business process by using one or more of the tools for continual improvement described in this chapter, it must ensure that the gains realized are secured and that the improved process constitutes the new baseline (or standard) for process execution. The establishment of this enhanced baseline entails the following steps.

8.3.1 Update Appropriate QMS Documentation

As stated in the previous section, appropriate QMS documentation should be revised once the decision is made to implement a permanent process change. The revised QMS documentation should be reviewed and approved prior to dissemination to all employees; minimally, the revised QMS documentation should be disseminated to all affected employees.

8.3.2 Deploy the Revised QMS Documentation

Once QMS documentation has been revised, the process of deploying the revised processes should begin. (Refer to Chapter 7.) This entails training employees on the process changes, and monitoring compliance to the revised processes. Compliance to the revised processes may be monitored in real time (as described in Chapter 7) and in internal quality audits. Appropriate corrective action should be taken when process execution is found to be noncompliant with the defined process — this includes determining whether additional employee training is necessary, or if additional process changes are necessary to facilitate process compliance.

REFERENCES AND BIBLIOGRAPHY

[ASQ99] ASQ and the Holmes Corp. 1999. *Quality 101 Self-Directed Learning Program*. Milwaukee: ASQ.

[Bas84] Basili, V. and Weiss, D. M. "A Methodology for Collecting Valid Software Engineering Data, IEEE Transactions on Software Engineering." Vol. 10, No. 3, Nov. 1984.

[Cro92] Crosby, P B. 1992. *Quality is Free: The Art of Making Quality Certain*. Mentor Books, New York.

[Dem86] Deming, W. E. 1986. *Out of the Crisis, MIT Center for Advanced Engineering Study*. Cambridge, MA.

[Fin00] Fine, D. L. and Read, W. L. "A Blueprint for Document Control: How to Develop, Maintain and Improve the System." *Quality Progress*, March 2000.

[Fos00] Foster, T. 2000. *Managing Quality: An Integrative Approach*. Upper Saddle River, N.J.: Prentice Hall.

[Gar01] Gardner, R. A. "Resolving the Process Paradox," *Quality Progress*, March 2001.

[Gar02] Gardner, R. A. "10 Process Improvement Lessons for Leaders." *Quality Progress*, Nov 2002.

[GoD01] Goetsch, D. L. and Davis, S. B. 2001. *Total Quality Handbook*. Upper Saddle River, N.J.: Prentice Hall,

[HoH01] Hoyer, R. W. and Hoyer, B. B. Y. "What is Quality?" *Quality Progress*. July 2001.

[Hoy94] Hoyle, D. 1994. *ISO 9000 Quality Systems Handbook*. Woburn, MA: Butterworth-Heinemann.

[ISO8402] *ISO 8402: Quality Management and Quality Assurance — Vocabulary*. ISO. 1994.

[ISO9000-1] *ISO Quality Management and Quality Assurance Standards, Part 1: Guidelines for Selection and Use, ISO 9000-1: 1994*. ISO. 1994.

[ISO9001] *ANSI/ISO/ASQ 9001-2000: Quality Management Systems — Requirements, International Organization for Standardization*. ISO. 2000.

[Jur88] Juran, J. M. 1988. *Juran's Quality Control Handbook, Fourth Ed*. New York: McGraw-Hill.

[Kap92] Kaplan, R. and Norton, D. "The Balanced Scorecard: Measures that Drive Performance," *Harvard Business Review*, Jan. 1992.

[Keating] Keating, E. K. et al. "Overcoming the Improvement Paradox." *European Management Journal*. Vol. 17, No. 2, pp. 120–134.

[Kee97] Keen, P. 1997. *The Process Edge: Creating Value Where it Counts.* Cambridge: Harvard Business School Press.

[Nan01] Nanda, V. *On Tailoring an Organizational Standard Software Development Process for Specific Projects, Proceedings of the Eleventh International Conference on Software Quality (ICSQ 11).* Pittsburgh, Oct. 2001.

[Nan03] Nanda, V. 2003. *ISO 9001:2000 — Achieving Compliance and Continuous Improvement in Software Development Companies.* Milwaukee: ASQ Quality Press.

[Nan97] Nanda, V. *A Methodical Approach to Establishing a Process Management Infrastructure in an Organization, Proceedings of the Software Technology and Engineering Practice Conference (STEP'97).* London. July 1997.

[Pag00] Page, S. B. "Research: The Key to Quality Policies and Procedures." *Quality Progress.* January 2000.

[P-CMM] Curtis, B. et al. 2001. *People Capability Maturity Model: Guidelines for Improving the Workforce.* Boston: Addison-Wesley.

[Qua02] Quality Glossary, *Quality Progress.* July 2002.

[Rom00] Romano, P. "ISO 9000: What Is its Impact on Performance?" *Quality Management Journal.* Vol. 7, Issue 3, 2000.

[RuB95] Rummler, G. A. and Brache, A. P. 1995. *Improving Performance: How to Manage the White Space on the Organization Chart.* San Francisco: Jossey-Bass Publishers.

APPENDICES

Appendix A

SAMPLE OUTLINE OF A PROJECT PLAN FOR QMS IMPLEMENTATION

1. Introduction
   - Purpose of this document <*Describe the purpose of this document and provide an overview of its content.*>
   - Terminology <*Provide definition of terms and acronyms used in this document.*>
2. Project Overview
   - Project goal <*Describe the goal of the QMS implementation project, so that the implementation goal meets the SMART criteria explained earlier in the book.*>
   - Scope of implementation <*Describe the scope of the QMS implementation in terms of the organizational functional areas that are in the scope of the implementation. Specify functional areas that are excluded from the implementation.*>
3. Project Team
   - Project organization <*Draw an organization chart of the project team. As an example refer to Figure 5.*>
   - Roles and responsibilities <*Describe the responsibilities of each role in the project organization, and any additional roles as appropriate.*>
4. Implementation Process
   - Implementation Strategy <*Describe the overall implementation strategy. (Refer to the related discussion in Chapter 2.)*>
   - Implementation phases and major milestones <*Describe the overall implementation process, including description of the planned*

stages of the project, beginning with initial planning and ending with assessment of the final implementation against project goals.>

5. Project Management
   - Project tracking and control mechanisms *<Provide an overview of the mechanisms that will be used to track progress and to coordinate and control the implementation. For example, describe different types of meetings of the implementation team that will be called to assess progress and to control project execution.>*
   - Progress reporting *<Describe when and how implementation progress will be reported to staff and senior management. Note that implementation progress typically is reported to senior management at periodic management reviews of the QMS.>*
6. Supplementary Information
   - Assumptions *<List assumptions made in establishing this project plan. Note that some assumptions may be such that if they are not satisfied, they may evolve into limitations or risks that threaten some or all of the project success factors.>*
   - Other project issues *<List any other issues pertaining to the implementation, such as project time-keeping code to be used by staff for reporting time spent on the project.>*
7. Related Documents
   - *<List other documents that are relevant to the QMS implementation. These include internal documents as well as industry documents.>*
8. Change History

Document Version	*Date*	*Change Performed By*	*Summary of Change*
<Identify each document version>	<Identify the date each document version was created>	<Identify who performed the change for each document version>	<Provide a brief summary of changes for each document version>

9. Appendices
   - Detailed Schedule *<Include the initial project schedule (preferably as a Gantt chart) showing a high-level overview of the major phases in the project, key activities in each phase, and dates.*

Provide a reference to where the current project schedule will be available (if it is not maintained as part of this document).>

- Risks *<List the initial project risks, probability of risk,* severity of risk,** risk exposure,*** risk mitigation, and risk contingency. Provide a reference to where the list of project risks will be continually maintained.>*
- Resource Requirements *<List resource requirements, such as consultants (in person-days or person-months), books, training needs, tools (e.g., software), professional memberships, and so on.>*

---

* Here, probability of risk is defined as the probability of occurrence of the risk — that is, the probability that the risk will evolve into a problem. This may be rated as high, medium, and low, each associated with a certain weight (in descending order).

** Severity of risk is defined as the severity of consequences if the risk evolves into a problem. This may be rated as high, medium, and low, each associated with a certain weight (in descending order).

***Risk exposure is obtained by multiplying the probability of risk with the severity of risk to arrive at a final score. Risks should be sorted in descending sequence by risk score, which yields the relative priority of the risks.

Appendix B

SAMPLE OUTLINE OF A QUALITY MANUAL*

Section 1: Introduction

- Purpose *<Describe the purpose of the Quality Manual.>*
- Scope *<Describe the scope of the QMS, including a high-level description of the business processes and office locations covered.>*
- Organizational Overview *<Provide a brief overview of the organization, its products, and its locations.>*

Section 2: QMS Overview

- Quality Policy *<Provide a statement of the organization's quality policy.>*
- Organization Chart *<Represent the organization's major functional areas (or departments) in an organization chart. Do not include names of employees in this organization chart, because such information is prone to change. Preferably, shade functional areas that are included in the scope of the QMS implementation.>*
- Roles and Responsibilities *<Provide a brief description of the role and key responsibilities of each functional area in the organization chart. Provide a brief, generic description of responsibilities for a typical process owner.>*
- Quality Management Representative *<Introduce the organization's quality management representative by name so that employees are*

* Reproduced with permission from Nanda, Vivek (Vic), *ISO 9001:2000 — Achieving Compliance and Continuous Improvement in Software Development Companies*, ASQ Quality Press, Milwaukee, 2003.

aware of the person in this key role. Briefly describe this person's responsibilities.>

- Management Review of the QMS *<State who (role) participates in the periodic management review of the QMS, what the periodicity of the management reviews is, what the quorum is, and the nature of items discussed at such reviews.>*

Section 3: QMS Structure

<Briefly describe the structure of the QMS in terms of types of QMS documentation and their hierarchy.>

Section 4: QMS Content

- Overview of the Metaprocess *<Provide an overview of the organization's business processes at the highest level of abstraction, and reference additional documentation, if any (such as process maps and procedures). Briefly describe functional areas that are outside the product development process but part of such a metaprocess, such as: Contracts & Legal departments; Sales; Marketing; Customer Support; and Customer Training. Reference QMS procedures, if any, that further describe these functional areas.>*
- Overview of the Product Development Process *<Introduce the three different types (or categories) of product development processes described below, and Key milestones in the product development process.>*
  - Management Processes *<Briefly describe the management processes — for example, management review of project progress, project management, and product management processes. Reference related QMS documentation, if any.>*
  - Technical processes *<Briefly describe the technical processes in the overall product development process. Reference related QMS documentation, if any.>*
  - Support Processes *<Briefly describe support processes of the product development process, such as quality assurance, employee training, etc. Reference related QMS documentation, if any.>*
  - 1. Milestone Reviews *<Identify the milestone reviews to be held. Reference related QMS documentation that specifies the milestone review criteria for each milestone, including the mechanism for handling of deviations.>*

Section 5: Quality Management System Standard to QMS Documentation Traceability Matrix (if applicable)

<Provide traceability between each requirements clause in the applicable quality management system standard and key QMS documentation.>

Section 6: Related Documents

<Provide a list of all QMS documents referenced in the Quality Manual, or otherwise related.>

Section 7: Document Change History

Document Version	*Date*	*Change Performed By*	*Summary of Change*
<Identify each document version>	*<Identify the date each document version was created>*	*<Identify who performed the change for each document version>*	*<Provide a brief summary of changes for each document version>*

Appendix C

SAMPLE QUALITY PROCEDURES

APPENDIX C.1: PROCEDURE FOR DOCUMENTING QMS PROCESSES

Section 1: Introduction

- Purpose
  This procedure describes the process to be followed for documenting QMS processes in procedures and work instructions.
- Scope
  This procedure describes all activities that must be executed once a need for a procedure or work instruction is identified. This includes researching the process to be documented, preparation of drafts of the documentation, reviews of the documentation with appropriate stakeholders, rework, approval, and release.
- Applicability
  - All departments in the organization must adhere to this procedure when documenting QMS processes.
  - Adherence to this procedure is recommended for creation of other types of QMS documentation, such as templates, forms, checklists, and other documents.
- Terminology
  None

Section 2: Process Flowchart

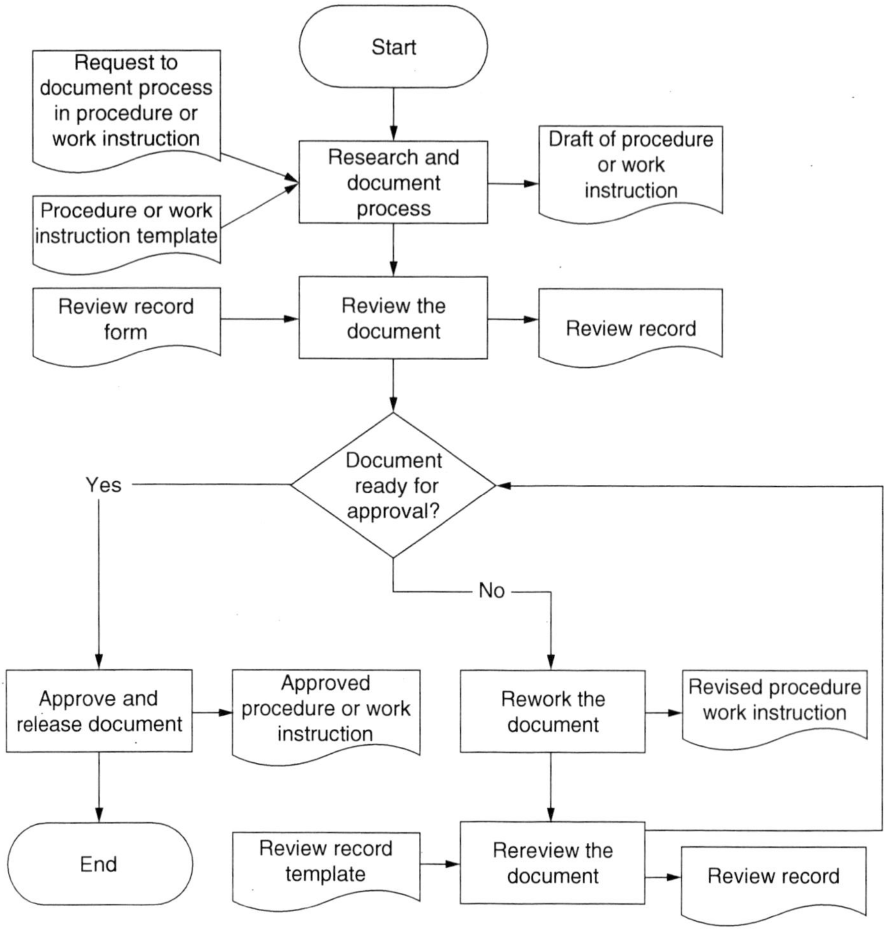

Figure 26 Flowchart for Creating Procedures and Work Instructions

Section 3: Entry Criteria

Request from process owner to create a procedure or work instruction.

Section 4: Inputs

Input	*Provider*
Request to create a procedure or work instruction	Process owner or quality management representative
Procedure or work instruction template	QMS documentation repository
Review record form	QMS documentation repository

Section 5: Activity Descriptions

Activity Description	*Output*	*Responsible*
1. **Research and document the process.** The author of the document researches the process to be documented and prepares an initial draft of the procedure or work instruction by using the appropriate template (see related documents [1] and [2]). Different techniques the author may use to research a process include process flowcharting, meetings or interviews with process practitioners to obtain process information, etc.	Draft of procedure or work instruction	Author
2. **Review the documented process.** The author provides the draft document for review to various stakeholders (in accordance with related document [3]). Generally, one or more informal reviews of a document are scheduled prior to a formal review. In some cases, a document may directly be scheduled for a formal review.	Review record	Review moderator

(continued)

Activity Description	*Output*	*Responsible*
(continued) The primary purpose of an informal review is to obtain feedback from the various stakeholders for the improvement of the document prior to its approval. The primary purpose of a formal review is to determine whether the document is ready to be formally approved, or if additional rework and rereviews are required. The review moderator formally logs all review comments made at the formal review by completing the review record form [4]. (In case of a walkthrough meeting, the review record is completed by the producer.) No review records are required for a desk check review.		
3. Document ready for approval? If the document is ready for approval, proceed to step 4; otherwise, proceed to step 5.		
4. *Approve and release document for use* If the document content is acceptable, initiate formal approval of the document. Once required approvals have been obtained, the document is submitted for storage in the controlled QMS documentation repository. (This constitutes release of the formally approved document.)	Approved document	Author
5. Rework the document If the document is not ready for approval, rework the document as per reviewer comments and resubmit it for review.	Revised document	Author
6. Rereview the document The reviewers rereview the document. If the document is deemed acceptable, it is approved and released; if not, additional rework and rereviews are performed as needed.	Review record	Review moderator

Section 6: Exit Criteria

The approved procedure or work instruction is submitted for storage in the QMS documentation repository.

Section 7: Outputs

Output	*Receiver*
Approved procedure or work instruction	QMS documentation repository
Review record	QMS documentation repository

Section 8: Process Metrics

Not applicable

Section 9: Related Documents

[1] Procedure template (Document number: QUA-TP-QMS-0001)
[2] Work instruction template (Document number: QUA-TP-QMS-0002)
[3] Document reviews procedure (Document number: QUA-PR-QMS-0005)
[4] Review record form (Document number: QUA-FR-QMS-0011)

Section 10: Document Change History

Document Version	*Date*	*Change Performed By*	*Summary of Change*
1.0	01-02-04	Vic Nanda	First approved version

APPENDIX C.2: PROCEDURE FOR HANDLING PROCESS IMPROVEMENT SUGGESTIONS

Section 1: Introduction

- Purpose
  This procedure describes the process to be followed for handling process improvement suggestions (including process change requests).
- Scope
  This procedure covers all activities in the life cycle of a process improvement suggestion, beginning with its initial submission and ending with its final disposition (closure).
- Applicability
  1. This process applies to the processing of:
  2. Process improvement suggestions received from employees

  Process improvement requests received from external sources, such as customers and external auditors

  Examples of process improvement requests to which this procedure applies include:
  - Lessons learned from product development projects (during project execution or from project postmortem analysis)
  - Improvement suggestion from employee
  - Need for a process change identified from root cause investigation of a reported product defect (or class of product defects)
  - Need for a process change identified from root cause investigation of a situation against which corrective and/or preventive action has been requested
  - Need for a process change identified during a quality audit
  - Need for a process change identified from metrics trends analysis
  - Request for a process change made by a customer or external auditor

  Note: This procedure applies regardless of whether the subject process is documented in a procedure (or other QMS documentation).*
- Terminology
  For the purpose of this procedure, a process change request may be regarded as the same as a process improvement suggestion.

---

* It is accepted that every process in the organization might not be documented, or different processes might be documented to different levels of detail. Nevertheless, a process improvement suggestion still may be submitted for a process. A change in the process may be reflected in updated process documentation or promulgated by training or communication to the affected employees.

Section 2: Process Flowchart

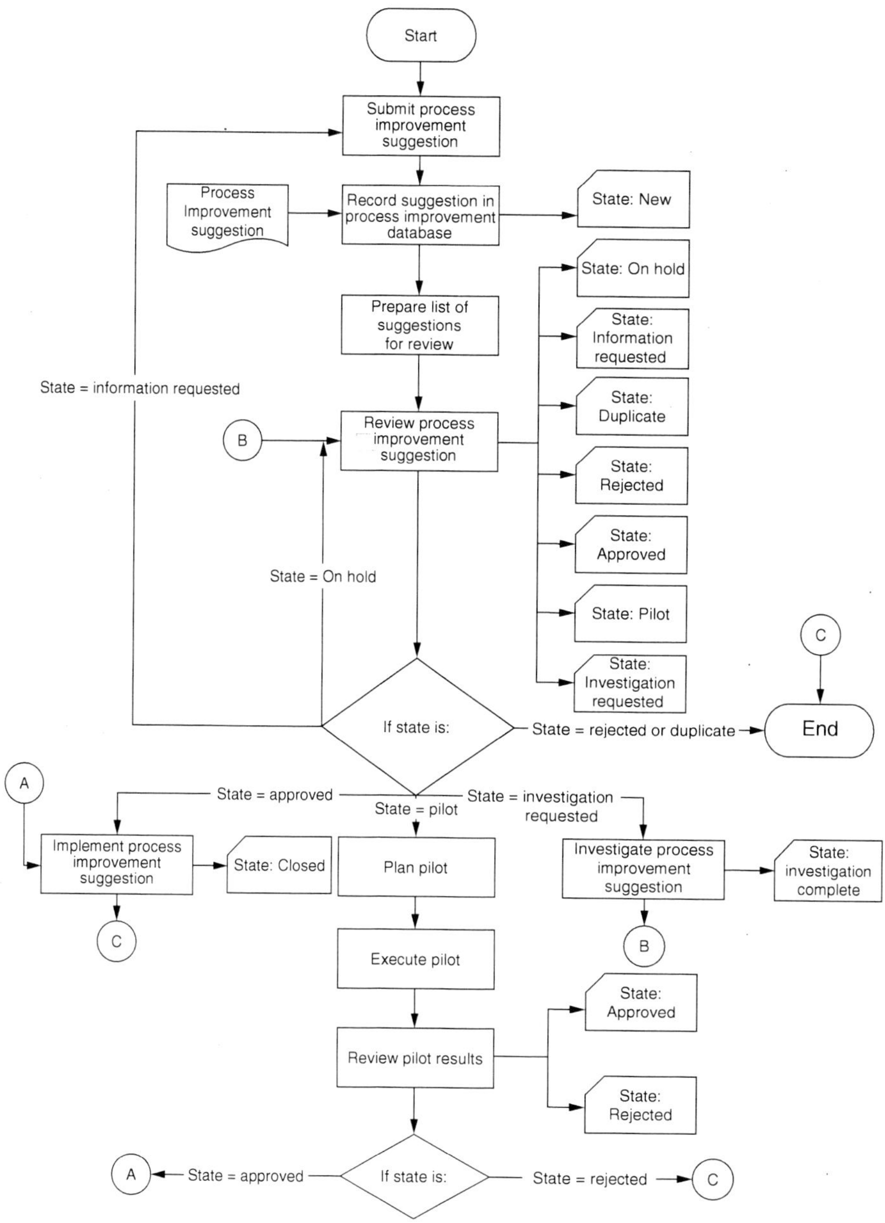

Figure 27 Flowchart for Handling of Process Improvement Suggestions

Section 3: Entry Criteria

Process improvement suggestion is received

Section 4: Inputs

Input	*Provider*
Process improvement suggestion	Employee or external source

Section 5: Activity Descriptions

Activity Description	*Output*	*Responsible*
1. **Submit process improvement suggestion.** A process improvement suggestion may be submitted by: • Employees by contacting a member of the quality assurance department (either in person, or by email); or • A member of the quality assurance department (including requests submitted on behalf of external sources).	Improvement suggestion	Employee
2. **Record suggestion in process improvement database** All process improvement suggestions are entered in the process improvement database by a member of the quality assurance department[a] (state of the suggestion at this time is "New"), and the submitter of the suggestion is informed that the suggestion has been formally logged and is provided the suggestion number for reference purposes.[b]	Improvement suggestion formally recorded in the process improvement database	Quality assurance department

(continued)

Activity Description	Output	Responsible
(continued)		
3. **Prepare list of suggestions for review.**		
Prior to each meeting of the process management council, the quality assurance department creates a report of all improvement suggestions in one of the following states: New, On hold, and Investigation complete. The quality assurance department reviews the report and has the authority to purge "on hold" suggestions: a. If the condition that caused the suggestion to be suspended is still true; or b. If the time period identified for the suspension of the suggestion has not yet expired. The list of suggestions then is provided to the process management council for review.	List of improvement suggestions for review at the next Process Management Council meeting	Quality assurance department
4. **Review process improvement suggestions**		
Process improvement suggestions are reviewed at the process management council meetings for anticipated benefit and cost of implementation. Each suggestion progresses to one of the following states: **On hold:** A suggestion may be put on hold if it is accepted as a valid suggestion, but there are conditions that prevent implementation of the suggestion in the near term. Suggestions put on hold should be annotated with an appropriate explanation and tagged with the date until which the suggestion is to remain suspended.	Disposition of process improvement suggestion	Process Management Council

(continued)

Activity Description	*Output*	*Responsible*
(continued) **Information requested:** A suggestion may be moved to the Information requested state if it is deemed incomplete or it requires further clarification from the submitter. **Duplicate:** A suggestion may be moved to the Duplicate state if it is similar to a previously submitted suggestion (in this case, a reference to the original submission must be recorded). **Rejected:** A suggestion may be rejected if the Process Management Council determines that the suggestion will be ineffective in achieving the stated improvement, is too expensive, or otherwise is unfeasible. In all cases, the rejection decision and reason for rejection must be communicated to the submitter of the suggestion. **Approved:** If a suggestion is Approved, the Process Management Council assigns it to an appropriate management person for implementation (typically the process owner), hereafter referred to as the suggestion owner. The submitter of the suggestion is notified regarding the approval decision (see step 6). **Pilot:** When appropriate, the Process Management Council may choose to pilot an improvement suggestion so that its results can be examined prior to making a final decision.		

(continued)

Activity Description	Output	Responsible
(continued)		
This is especially true in the case of complex changes; expected widespread impact of a change; highly innovative changes; or perceived high risk of a flawed implementation. If a suggestion is to be piloted, the Process Management Council assigns it to an appropriate management person (suggestion owner) (see Step 7).		
Investigation requested: In certain situations, the Process Management Council may request further investigation by a suitable subject matter expert. Results of the investigation are then considered by the Process Management Council in making a final decision (see step 5).		
5. **Investigate process improvement suggestion**		
If the Process Management Council requests further investigation, the assigned subject matter expert analyzes the suggestion for feasibility of implementation (including cost and time), implementation approach, and anticipated benefits. The subject matter expert then makes a recommendation on the acceptance or rejection of the suggestion, and whether or not it should be piloted (if acceptance is recommended).	Recommendation regarding acceptance or rejection of suggestion	Subject matter expert (designated by the Process Management Council)

(continued)

Activity Description	Output	Responsible
(continued)		
6. Implement process improvement suggestion		
The identified suggestion owner implements the improvement suggestion. The improvement suggestion is formally closed by the Process Management Council and the submitter of the suggestion is informed about the closure.	Implementation of the suggestion is complete	Suggestion owner Process Management Council
7. Plan pilot		
When necessary, the Process Management Council may request a suggestion to be piloted. The following items should be addressed in planning a pilot: scope of the pilot; participants in the pilot; duration of the pilot; and desired results.	Pilot plan	Suggestion Owner
8. Execute pilot		
The suggestion owner coordinates the execution of the pilot in accordance with the pilot plan.	Recommendation by suggestion owner after pilot completion	Suggestion owner
9. Review pilot results		
The Process Management Council reviews the pilot results and recommendation from the pilot, and makes a final decision (approval or rejection) regarding the improvement suggestion.	Disposition of process improvement suggestion	Process Management Council

[a] Alternatively, a company may choose to grant such privilege to all employees.
[b] It is preferable that such a notification be automatically sent by the process improvement database tool.

Section 6: Exit Criteria

Disposition of the process improvement suggestion by the Process Management Council

Section 7: Outputs

Output	Receiver
Audit trail and final disposition decision (approval or rejection) for each process improvement suggestion	Submitter and suggestion owner (in case of approval)
Process improvement or change (for approved suggestions)	NA

Section 8: Process Metrics

Turnaround time for implementation of process improvement suggestions

Section 9: Related Documents

Process Improvement Suggestion Submission Form (electronic form in Process Improvement Database)

Section 10: Document Change History

Document Version	Date	Change Performed By	Summary of Change
1.0	10-04-03	Vic Nanda	First approved version

APPENDIX C.3: SUPPLIER MANAGEMENT AND CONTROL PROCEDURE

Section 1: Introduction

- Purpose
  This procedure describes the process to be followed for selecting new suppliers and for managing and controlling supplier performance.
- Scope
  This procedure describes the process for evaluating and selecting a new supplier based on the supplier's ability to meet organizational requirements. It also describes mechanisms for managing, controlling, and improving supplier performance so that organizational requirements continue to be met.
- Applicability
  - This process applies to the purchase of:
    - All products used in the design and development of the organization's products; and
    - All products delivered as part of (or for use with) the final product(s).
  - This process does not apply to the purchase of:
    - Capital equipment, such as personal computers, laptops, printers, fax machines, and other such office IT equipment
    - Office supplies, such as office furniture and stationery
- Terminology
  None

Section 2: Process Flowchart

Not applicable

Section 3: Entry Criteria

1. Request to select a new supplier is received from a relevant authority within the organization; or
2. Request to procure a new product for use in the product development process, or for use in (or with) the finished product, is received from a relevant authority within the organization.

Section 4: Inputs

Input	Provider
Request to select a new supplier	Management personnel from relevant department
Request to procure a new product	Management personnel from relevant department

Section 5: Activity Descriptions

Activity Description	Output	Responsible
1. **Supplier selection** A prospective supplier must be deemed satisfactory after evaluation at each step in the selection process in order to be considered any further in the selection process. If a supplier is deemed unsatisfactory at any step in this selection process, any further evaluation of the supplier requires authorization from the quality management representative. All quality records produced during the supplier selection process, such as completed questionnaire responses and supplier assessment reports, are maintained by the quality assurance department (in accordance with related document [1]).		

(continued)

Activity Description	*Output*	*Responsible*
(continued)		
1.1 Quality questionnaire		
At the start of the supplier selection process, the quality assurance department sends a quality questionnaire [2] to all prospective suppliers with a request that it be completed and returned within two weeks.	Decision on whether or not supplier evaluation should proceed to next step	Quality assurance and purchasing departments
The purpose of this questionnaire is to obtain pertinent information about each supplier (including supplier history, product price, production facilities and capacity utilization, reputation, and references) and the supplier's products and QMS.		
The purchasing and quality assurance departments jointly review the completed quality questionnaires received from all suppliers to ascertain whether or not each supplier has the ability to meet organizational requirements, and if evaluation of the supplier may proceed to the next step.		
1.2 Onsite quality assessment		
The second step involves an onsite assessment of the supplier's QMS by the quality assurance department. This assessment is essential for distinguishing between the maturity of the QMSs of different prospective suppliers — information that is critical for selecting the best supplier.		
Each supplier is evaluated against a set of quality requirements and/or criteria documented in an assessment checklist created by the quality	Onsite quality assessment checklist	Quality assurance department

(continued)

Activity Description	Output	Responsible
(continued) assurance department. Inpreparing the checklist for the onsite assessment, the quality assurance department may adopt one of the following approaches: • It may choose to use requirements from only a relevant industry QMS standard;[a] or • In addition to requirements from a relevant industry QMS standard, it may use additional requirements/criteria that have been established internally by the organization; or • It may use solely internally specified requirements/criteria.		
After the assessment, the quality assurance department produces a formal assessment report that lists all assessment findings, and sends the report to the prospective supplier. The prospective supplier is given two weeks to provide a corrective action plan for the findings noted in the report.	Supplier assessment report	Quality assurance department
Based on the results of the supplier onsite quality assessment (and the response of the prospective supplier), the quality assurance department makes a recommendation on supplier selection to the purchasing department.		

(continued)

Activity Description	Output	Responsible
(continued)		
1.3 Supplier selection decision		
At this stage, if appropriate, the purchasing department may decide to place a trial order to obtain direct evidence of the supplier's ability to meet organizational requirements. During the trial period, the purchasing and quality assurance departments closely monitor supplier performance, which includes delivery performance and quality of products received. (This is verified by the receiving personnel by inspection and/or testing of the products against requirements.)		
The purchasing and quality assurance departments jointly evaluate the supplier based on the results of the onsite assessment and experience from the trial order (if one was placed) and make a final decision regarding supplier selection. In case of disagreement between the quality assurance and purchasing departments regarding supplier selection, the issue may be escalated to line management of both organizations, or to the QMS steering group.	Supplier selection decision	Purchasing and quality assurance departments
Once a supplier has been selected, the purchasing department adds it to the official approved supplier list of all approved suppliers of the organization, along with their contact information.		

(continued)

Activity Description	*Output*	*Responsible*
(continued)		
No product may be purchased from a supplier not on the approved supplier list (unless otherwise authorized by the quality management representative).	Updated approved supplier list	Purchasing department
2. Supplier monitoring and control After a supplier has been selected and added to the approved supplier list, an ongoing supplier management and control program is initiated to monitor supplier performance against organizational expectations and requirements, mitigate supplier-related risks, and provide input to the supplier for continuous improvement and prioritization of quality improvement tasks. Different means for supplier management and control include:		
2.1 Approved parts list When a supplier provides products (or parts) for inclusion in the final end product, control is exercised on the supplier to preclude the use of inferior parts (or components). This is necessary to ensure consistency in the quality of the supplied products (or parts), and to preclude the use of inferior material to cut costs or for other reasons. In order to ensure use of only agreed-upon parts by the suppliers, the purchasing department and the supplier establish an approved parts list.[b] A joint review of the parts list is held to cover such issues as appropriateness of the parts for the intended use and feasibility of easily sourcing the parts.	Approved parts list	Purchasing department

(continued)

Activity Description	*Output*	*Responsible*
(continued)		
2.2 Incoming inspection and testing of received goods In addition to specifying the parts approved for use by the supplier, it is verified that the supplier is using only approved parts. This is accomplished by performing incoming inspection (and testing, where appropriate) of the received goods (in accordance with related document [3]) and/or by performing outgoing inspection at the supplier site as per the contractual agreement with the supplier. (This is in addition to inspection and test performed by the supplier.) It is required that incoming inspection cover 100% of the received goods (i.e., full coverage), unless deviation to this rule is authorized by the quality management representative. Outgoing inspection performed at the supplier site might be limited to a random sampling as per the discretion of the quality assurance personnel executing the task.	Quality records indicating one of the following dispositions of the received goods: approved, rejected, or rework	Receiving personnel
2.3 Supplier performance data The purchasing department continually tracks the performance of all suppliers by using appropriate quality metrics.[c] If a quality metric requires that measurement data be collected by the supplier and by the organization, the quality metric is defined in consultation with the supplier. All suppliers are	Supplier performance data	Purchasing department

(continued)

Activity Description	Output	Responsible
(continued) required to report their data periodically for the agreed-upon quality metrics to the organization, as needed. When necessary, such an obligation is recorded in the contractual agreement with the supplier.		
The purchasing and quality assurance departments jointly review supplier performance data each month, and appropriate action is requested from the supplier by means of a formal corrective and/or preventive action request, as needed. The purchasing department maintains supplier performance data for all suppliers.		
The quality management representative is responsible for presenting a summary of supplier performance data at the periodic management review meetings to ensure continued management awareness of supplier performance, and for discussion of related quality concerns.		
2.4 Annual surveillance audits		
To verify that each supplier is continuing to execute its processes in accordance with its defined QMS, and to assess the supplier's continued ability to meet organizational requirements, each supplier is required to undergo an annual surveillance audit conducted by the quality assurance department. For a supplier to continue to maintain its status as an approved supplier, it must	Records of annual surveillance audit of supplier	Quality assurance department

(continued)

Activity Description	*Output*	*Responsible*
(continued) pass this annual audit and agree to address the findings from the audit. The quality assurance department maintains records of the annual surveillance audit for each supplier.		
2.5 Supplier disqualification		
In certain situations, it might be necessary to remove a supplier from the approved supplier list. A supplier may be disqualified by consensus of the purchasing and quality assurance departments; disagreements between the two departments may be escalated to respective line management.	Removal of supplier from approved supplier list	Quality assurance and purchasing departments
Supplier disqualification may be warranted if the supplier fails the annual surveillance audit; the supplier repeatedly ignores requests for corrective/preventive action, provides an unsatisfactory response to such a request, or deliberately takes ineffective action in response to such a request; or if there is continuous subpar supplier performance (as evidenced by the performance metrics) or failure to meet organizational requirements. This includes technical requirements, quality requirements, and on-time delivery requirements.		

[a] An example of a supplier evaluation matrix that uses an assessment checklist based solely on quality requirements in the ISO 9001:2000 standard is shown in Annex A. This and the subsequent annexes to this procedure are for explanation purposes only and should not be regarded as annexes of this procedure per se.

[b] An example of an approved parts list is shown in Annex B.

[c] Examples of quality metrics that may be used to monitor supplier performance are shown in Annex C.

Section 6: Exit Criteria

Decision to withdraw a supplier from the approved supplier list, either due to formal supplier disqualification or for other business reasons

Section 7: Outputs

Output	*Receiver*
Onsite quality assessment checklist	QMS documentation repository
Supplier assessment report (including completed quality questionnaire and results of onsite assessment)	Quality assurance department records
Supplier performance data	Purchasing department records
Approved supplier list	Purchasing department records
Approved parts list	Purchasing department records
Records of annual surveillance audit of suppliers	Quality assurance department records

Section 8: Process Metrics

- Defect detection rate
- On-time delivery performance
- Defect types

Section 9: Related Documents

[1] Control of quality records procedure (Document number: QUA-PR-QMS-0017)

[2] New Supplier Selection — Quality Questionnaire (Document number: QUA-CHK-QMS-0024)

[3] Receiving inspection and test procedure (Document number: REC-PR-QMS-0022)

Section 10: Document Change History

Document Version	*Date*	*Change Performed By*	*Summary of Change*
1.0	03-17-04	Vic Nanda	First approved version

Section 11: Annexes

Annex A

This annex provides an example of a supplier evaluation matrix that may be used for comparing two or more suppliers during the supplier selection process. The evaluation matrix shown contains supplier selection quality requirements (criteria) derived solely from the well-known ISO 9001:2000 QMS standard.

Let us now examine the supplier evaluation matrix in detail:

The matrix includes ISO 9001:2000 requirements only to the "x.y" level. Each requirement being assessed is given a weight (according to importance), which can be multiplied by the associated score to arrive at the element score. The auditing organization assigns the requirement weight based on how important the organization regards each requirement. The four-point scoring scale awards points based on whether the assessed supplier is best in class (score = 3), meets requirement (score = 2), partially meets the requirement (score = 1), or fails to meet the requirement (score = 0). The maximum possible score of 180 represents the score an organization can achieve if it is assessed as best in class (score = 3) for each requirement. An organization may establish a minimum "overall rating" that all its potential suppliers must achieve before further discussion regarding the supplier's financial and business viability can be held.

In the example shown (Table 17), assume that suppliers 1 and 2 are ISO 9000–registered, while supplier 3 is not. Also assume that the organization has established a minimum acceptance score of 110 for a supplier to be considered further in the selection process. In this example, assessment results show that although both the suppliers are ISO 9000–registered, supplier 1 is significantly superior to supplier 2. Supplier 3 fails to meet the minimum required score of 110 to stay in contention for selection. Therefore, at this point, the quality assurance department may recommend supplier 1 for selection, or both suppliers 1 and 2 may be evaluated further for financial and business viability before a final decision is made.

Annex B

This annex shows an example of an approved parts list (Table 18). Each part is identified with a unique part number, along with the primary vendor ("P") and secondary vendor ("S") from whom the part may be procured. Specifying the part vendor prevents the supplier from sourcing the parts from questionable vendors, and provides a high degree of assurance to the organization that its supplier will deliver products of consistent quality. A separate parts list for each purchased product should be maintained. Changes to the approved parts list may be performed only upon consultation between the organization and the associated suppliers. In addition, when appropriate, a mechanism should be established for handling of occasional deviations from the approved parts list. This is needed to accommodate situations where it might not be possible to procure a part from an authorized vendor, and the part might need to be substituted for an equivalent part procured from a vendor not on the approved parts list.

Annex C

This annex provides examples of quality metrics that may be used to monitor supplier performance.

Defect detection rate: This metric entails recording of monthly defect data by the organization to track the percentage of defective products received from the supplier. (Refer to Figure 28.)

This chart shows the percentage of defective pieces of product "Y" supplied by supplier X." In January 2003, 10% of the entire quantity of product "Y" was found to be defective during incoming inspection. In September 2003, the supplier instituted new quality control practices to reduce the unacceptably high defect rate. These efforts were marginally successful, as evidenced by the decrease in defect rates from September to December 2003.

On-time delivery performance: This metric entails recording the percentage of total deliveries of a product by the supplier that were received on time; percentage of total deliveries received a week late; percentage of total deliveries received two weeks late; and so on. An example of an on-time delivery performance chart is provided in Figure 29.

This shows the percentage of on-time deliveries of product "Y" from supplier "X" for each quarter of the years 2001 and 2002. In the first quarter of 2001, 65% of the deliveries were on time, 15% were up to a week late, 10% were more than a week but less than 2 weeks late, and 10% were more than two weeks late.

Defect types: This metric entails categorizing all the defects found in incoming inspection of a received product into separate defect types. (See Figure 30.) This enables an organization to identify the most commonly occurring defect types and work with its supplier to systematically minimize and eliminate them.

The chart in Figure 30 shows the distribution of defects by type. Such defect type data may be maintained internally by the supplier to record the types of defects found during the manufacturing process, as well as by the organization to track the types of defects found during incoming inspection of received products. The chart above shows only the defects found by

Table 17 Supplier Evaluation Matrix

ISO 9001:2000 Clause	*ID*	*Quality Requirement*	*Weight*	*Supplier 1*		*Supplier 2*		*Supplier 3*	
			1 to 3	*Score 0 to 3*	*Element Score*	*Score 0 to 3*	*Element Score*	*Score 0 to 3*	*Element Score*
		Please rate the potential supplier for each of the ISO 9001:2000 requirements	*1 Minor 2 Major 3 Critical*	*0 Fails req. 1 Partially met 2 Meets req. 3 Best in class*		*0 Fails req. 1 Partially met 2 Meets req. 3 Best in class*		*0 Fails req. 1 Partially met 2 Meets req. 3 Best in class*	
Quality Management System	4.1	General requirements	3	2	6	2	6	0	0
	4.2	Documentation requirements	3	2	6	2	6	1	3
		Total			12		12		3
Management Responsibility	5.1	Management commitment	3	3	9	2	6	0	0
	5.2	Customer focus	3	3	9	2	6	0	0
	5.3	Quality policy	3	2	6	2	6	2	6
	5.4	Planning	3	3	9	2	6	0	0
	5.5	Responsibility, authority, and communication	2	2	4	2	4	1	2
	5.6	Management review	3	2	6	2	6	1	3
		Total			43		34		11
Resource Management	6.1	Provision of resources	3	2	6	2	6	1	3
	6.2	Human resources	3	2	6	3	9	1	3
	6.3	Infrastructure	2	2	4	2	4	2	4
	6.4	Work environment	2	2	4	2	4	2	4
		Total			20		23		14

ISO 9001:2000 Clause	*ID*	*Quality Requirement*	*Weight*	*Supplier 1*		*Supplier 2*		*Supplier 3*	
			1 to 3	*Score 0 to 3*	*Element Score*	*Score 0 to 3*	*Element Score*	*Score 0 to 3*	*Element Score*
		Please rate the potential supplier for each of the ISO 9001:2000 requirements	*1 Minor* *2 Major* *3 Critical*	*0 Fails req.* *1 Partially met* *2 Meets req.* *3 Best in class*		*0 Fails req.* *1 Partially met* *2 Meets req.* *3 Best in class*		*0 Fails req.* *1 Partiallymet* *2 Meets req.* *3 Best in class*	
Product Realization	7.1	Planning of product realization	2	2	4	2	4	1	2
	7.2	Customer-related processes	3	2	6	2	6	1	3
	7.3	Design and development	3	3	9	2	6	1	3
	7.4	Purchasing	2	2	4	2	4	1	2
	7.5	Production and service provision	2	2	4	2	4	2	4
	7.6	Control of monitoring and measuring devices	2	2	4	2	4	2	4
		Total			31		28		18
Measurement, analysis, and improvement	8.1	General	2	2	4	2	4	1	2
	8.2	Monitoring and measurement	3	2	6	2	6	0	0
	8.3	Control of nonconforming product	2	2	4	2	4	2	4
	8.4	Analysis of data	3	2	6	2	6	0	0
	8.5	Improvement	3	2	6	2	6	1	3
		Total			26		26		9
		Overall Rating (Maximum Score: 180, Minimum Required: 110)	60		132		123		55

Table 18 Approved Parts List (partially complete)

ABC Corporation Approved Parts List

Document Owner: Purchasing department — Author: Bill Lehto
Document Number: PR-APL-1-0002 — Version: 1.2 — Last Updated: 11/30/01

Approved Parts for Model Z0X Passenger Car Seats

Part Number	*Description*	*Part Vendors*
SUBS1000191	Premium black leather seat cover	Hides Impex, Inc. (P) Covers-r-Us (S)
SUBS1010799	Head support mechanism	Moulds America Plc. (P) Smart Backs, Inc. (S)
SUBS2000198	Seat foam	Foamers, Inc. (P) Seats Unlimited, Inc. (S)

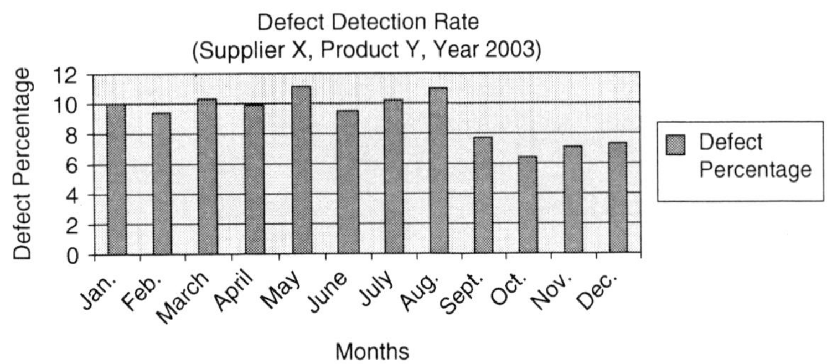

Figure 28 Defect Detection Rate

the organization during incoming inspection. The chart shows that during the year 2003, deliveries from supplier X had approximately 30 to 45% type "A" defects, 50 to 60% type "B" defects, 3 to 11% type "C" defects, and 2 to 7% type "D" defects. To achieve the greatest reduction in defects, and hence the greatest return on the investment, it will be desirable to investigate the root cause of defects of type "B."

For each performance metric, the organization and its suppliers should agree upon goals to foster continuous improvement in the items being measured. The organization also may make recommendations to its suppliers regarding suggested quality practices that will help achieve the established goals.

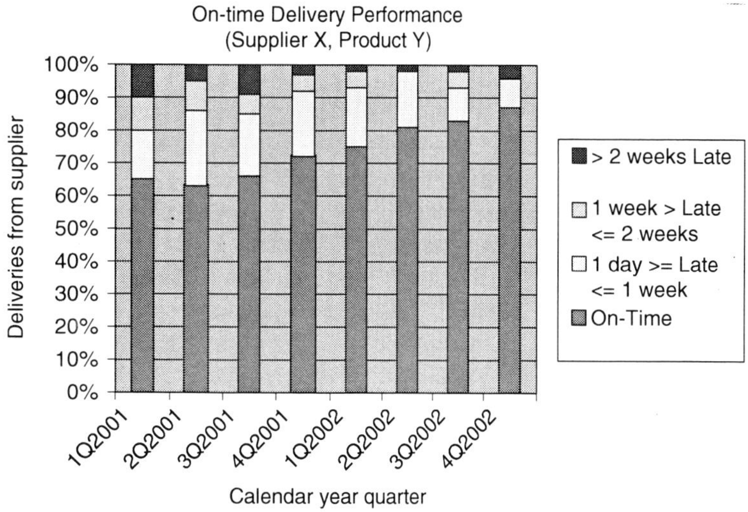

Figure 29 On-time Delivery Performance

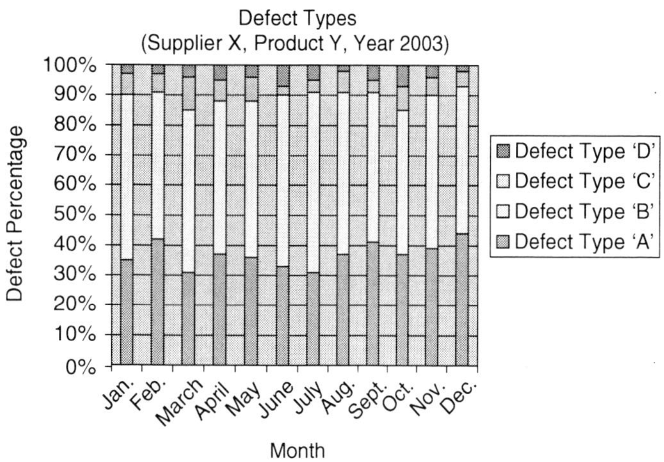

Figure 30 Defect Types

APPENDIX C.4: FORMAL AND INFORMAL REVIEWS PROCEDURE

Section 1: Introduction

- Purpose
  This procedure describes the process to be followed for performing formal and informal reviews of artifacts produced during product development and support.
- Scope
  This procedure describes different types of reviews (formal and informal), and when to conduct each. The procedure also describes the different roles involved in formal reviews, and provides detailed guidance on the process for each type of review.
- Applicability
  - This procedure applies to all employees involved in product development and support. This procedure applies to the review of the project/product documentation, the product and its components, and the QMS documentation (such as procedures).
  - This procedure may be used, as appropriate, for review of other company documentation such as contractual agreements, marketing documentation, financial documentation, and other business-related documents.
- Terminology
  - Desk check: A desk check is an informal review in which a person or a group privately reviews a deliverable to provide feedback to the producer. A desk check is performed privately by each reviewer and the reviewer directly provides comments to the producer.
  - Walkthrough: A walkthrough is a formal meeting in which the producer of an artifact leads the reviewers through the review of an artifact in a systematic manner. The producer provides information on and explanations of the artifact to the reviewers, while they ask questions and provide interactive feedback. The producer creates a review record to capture the action items arising from the reviewers' comments. A walkthrough meeting usually is preceded by a desk check.
  - Inspection: An inspection is a formal review meeting in which an artifact is inspected by an inspection team with well-defined roles (producer, moderator, recorder, and inspector). The moderator first clarifies the inspection approach; for example, a page-by-page or round-robin technique for a document review. Unlike a walkthrough, all discussions are kept to a minimum, and the

sole objective is to maximize discovery of defects, not to brainstorm solutions or alternate approaches. (These are usually handled outside the review meeting.) The recorder creates a review record to capture the action items arising from the reviewers' comments. If an inspection meeting is held, it generally is preceded by a desk check.

- Moderator: A moderator is a person who is requested by an artifact producer to conduct and control an inspection meeting*. The moderator assists the producer in review preparation by distributing the review material and notifying reviewers of the date, time, and place of the review. At the start of the review, the moderator determines if essential reviewers are present for an adequate review, recommending rescheduling of the review when necessary. The moderator controls the conduct of the review and ensures that a review record is created. The moderator also may play the role of a reviewer. After the review, the moderator verifies that all action items arising from the review are addressed adequately and determines if additional rework and/or rereview is necessary. The moderator distributes the revised review material to the reviewers prior to initiation of formal approval by the producer.
- Reviewer: A reviewer of an artifact is a person who is requested by the artifact producer to provide review feedback on the artifact. A reviewer is responsible for reviewing the artifact and providing comments to the producer in private (for a desk check) or in a review meeting (for a walkthrough or inspection).
- Recorder: A recorder is assigned (as needed) by the moderator during a walkthrough or inspection meeting to take meeting notes and complete a review record for the meeting. Alternately, a moderator may decide not to have a separate recorder and may take meeting notes and complete the review record himself.

* Although a moderator is not generally required for a walkthrough meeting, the producer may request one.

Section 2: Process Flowchart

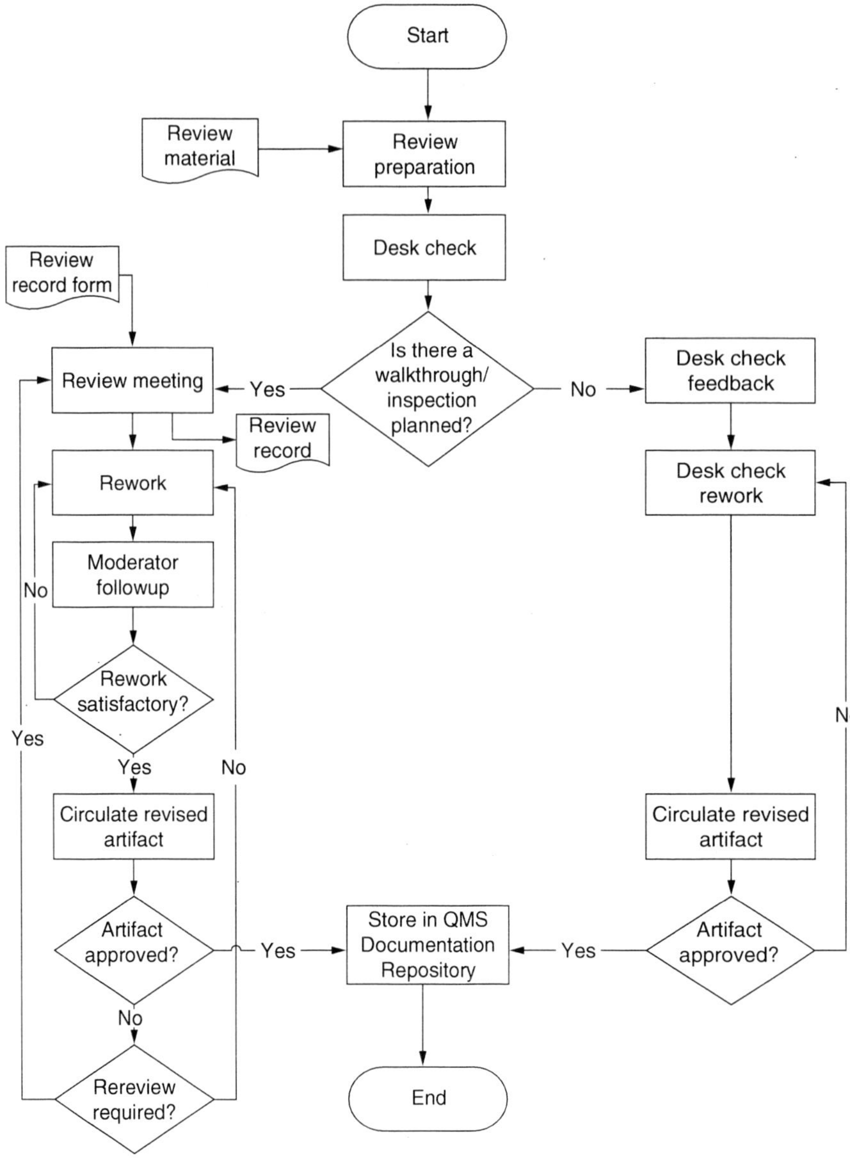

Figure 31 Flowchart for Formal and Informal Reviews

Section 3: Entry Criteria

Availability of review material

Section 4: Inputs

Input	*Provider*
Review material	Document author
Review record form (for walkthrough and inspection only)	QMS documentation repository

Section 5: Activity Descriptions

Formal and informal reviews enable an organization to improve the quality of its artifacts by facilitating detection of:

- Defects
- Missing information in documents (or missing parts in products)
- Use of inferior quality or incorrect parts
- Poor workmanship
- Technical deficiencies in documents
- Noncompliance with requirements (validation)
- Violation of applicable standards

Activity Description	*Output*	*Responsible*
1. **Review preparation** The producer (author) distributes the review material to all the reviewers for desk check and requests their comments by a specified date. Unless more time is necessary, reviewers are provided one week to prepare for a review meeting. If a walkthrough or inspection is planned, the review material is distributed by the moderator. The reviewers also are informed about the date, time, and place of the review meeting.	Review material Review date, time, and place	Producer or moderator (as applicable)
2. **Desk check** The reviewers review the material in private.	Feedback comments	Reviewers

(continued)

Activity Description	*Output*	*Responsible*
(continued)		
3. Is there a walkthrough/inspection planned?		
If a walkthrough/inspection is planned, proceed to step 4; otherwise proceed to step 5.	Walkthrough/inspection decision	Producer[a]
4. A walkthrough/inspection is planned		
4.1 Review meeting		
The review meeting (walkthrough or inspection) is conducted in accordance with the descriptions provided earlier in this procedure. A review record is produced for the meeting by completing the review record form [1].	Review record	Review participants
4.2 Rework		
The producer reworks the artifact in accordance with the action items in the review record and by the rework due date agreed to in the meeting. The producer then submits the revised artifact to the moderator for verification.	Revised artifact	Producer
4.3 Moderator follow-up		
The moderator verifies that the rework was performed in accordance with the review record.	Revised artifact (verified)	Moderator
4.4 Rework satisfactory?		
If the rework is deemed unsatisfactory, the artifact is returned to the producer for additional rework, as needed. If the rework is deemed satisfactory, the process proceeds to the next step.	Rework decision	Moderator
4.5 Circulate revised material		
The moderator provides the revised artifact to the reviewers to determine acceptability for formal approval.		Moderator
4.6 Artifact approved?		
If artifact is approved, process proceeds to step 6; otherwise, a determination is made whether additional rework, or additional rework and rereview, is required.	Approval decision	Approvers

(continued)

Activity Description	Output	Responsible
(continued)		
4.7 Rereview required? If rereview is required, one is held (step 4.1); if not, additional rework is performed (step 4.2).	Rereview decision	Moderator
5. **A walkthrough/inspection is not planned**		
5.1 Desk check feedback If a walkthrough is not planned, reviewers provide desk check feedback directly to the author (by e-mail, by returning marked-up copy with comments, or in person).	Review comments	Reviewers
5.2 Desk check rework The producer reworks the artifact in accordance with the review comments received. The producer and the respective reviewer discuss areas of disagreement.	Revised artifact	Producer
5.3 Circulate revised material The producer provides the revised artifact to the reviewers to determine acceptability for formal approval.		Producer
5.4 Artifact approved? If the artifact is approved, the process proceeds to step 6; otherwise, additional rework is performed (step 5.2).	Approval decision	Approvers
6. **Store in QMS documentation repository** If the artifact is approved, the producer submits the approved artifact and associated review records (if any) for storage in the QMS documentation repository.		Producer

[a] When appropriate, this decision may be made by the author, unless the QMS requires a particular type of review of the artifact.

Section 6: Exit Criteria

- Approved review material
- Completed review record (for walkthroughs and inspections) submitted for storage in controlled location.

Section 7: Outputs

Output	*Receiver*
Approved artifact	QMS documentation repository
Completed review record (for walkthrough and inspection only)	QMS documentation repository

Section 8: Process Metrics

Number of major remarks made at each formal review

Section 9: Related Documents

[1] Review record form (Document number: QUA-FR-QMS-0011)

Section 10: Document Change History

Document Version	*Date*	*Change Performed By*	*Summary of Change*
1.0	04-19-04	Hersh Rahul Nanda	First approved version

APPENDIX C.5: CORRECTIVE AND PREVENTIVE ACTION PROCEDURE

Section 1: Introduction

- Purpose
  This procedure describes the process followed by the quality assurance department for initiating, monitoring, and following up on corrective and preventive action requests (CARs and PARs).
- Scope
  This procedure covers all activities, beginning with the initiation of a corrective or preventive action request and ending with the closure of the request.
- Applicability
  This procedure applies to situations observed by or reported to the quality assurance department personnel during normal business operations that are deemed to require corrective or preventive action.

This procedure also may be used by the quality assurance department to request corrective or preventive actions as a result of situations brought to its attention from external sources, such as customer complaints, customer audits, or other external audits and assessments.

This procedure does not apply to observed products defects that require corrective action, or requests for preventive action pertaining directly to the product. Product-related corrective and preventive action requests are addressed in [3].

Terminology*

Corrective action: The implementation of a solution that results in the reduction or elimination of a nonconformance.

Preventive action: Action taken to remove or improve a process to prevent potential future occurrences of a nonconformance.

* These definitions are taken from [Qua02] with permission.

Section 2: Process Flowchart

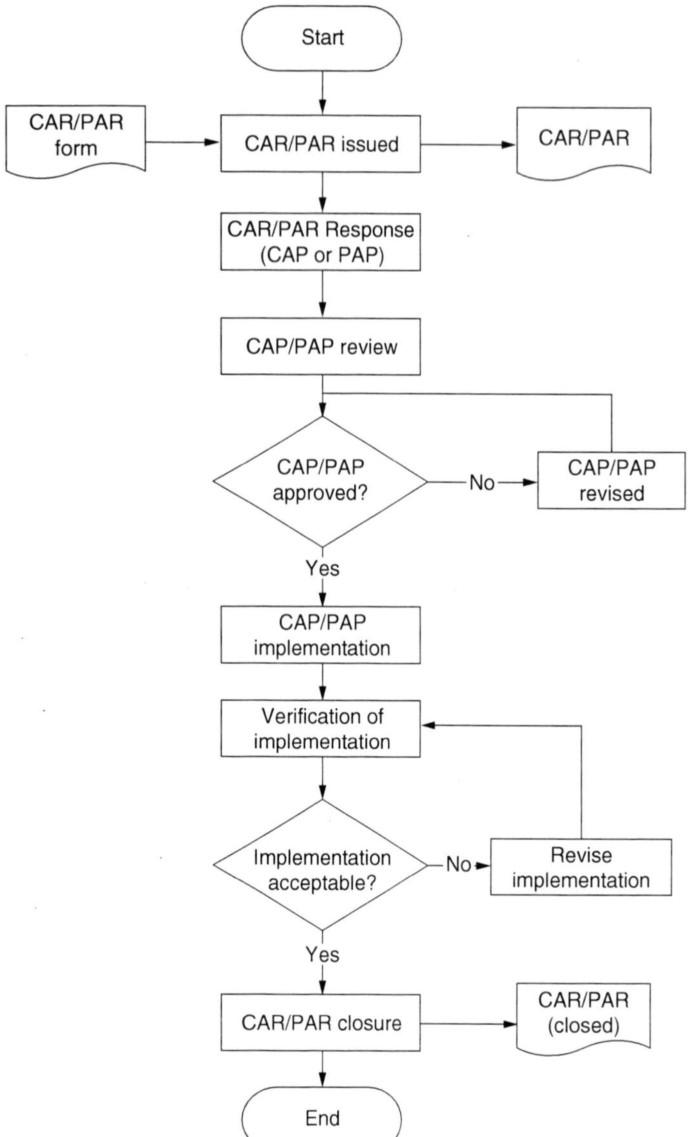

Figure 32 Flowchart for Corrective/Preventive Action Requests

Section 3: Entry Criteria

An observed condition that requires corrective or preventive action

Section 4: Inputs

Input	*Provider*
Actionable information on condition requiring corrective or preventive action	Member of quality assurance department
CAR/PAR form	QMS documentation repository

Section 5: Activity Descriptions

Activity Description	*Output*	*Responsible*
1. **CAR/PAR issued** The quality assurance department issues a CAR/PAR for an observed condition requiring corrective or preventive action. The originator of the CAR/PAR completes appropriate sections in the applicable form [1, 2]. Conditions that may warrant a CAR or PAR, as appropriate, include: • Violation of defined processes, procedures, plans, or applicable standards • Inadequate or ineffective processes (or practices) or other incidents that are known to have an undesirable effect on product quality, service quality, or customer satisfaction • Existence of conditions that may lead to an undesirable affect on product quality, service quality, or customer satisfaction • Reported defect or deficiency in product or service	CAR or PAR (as appropriate)	Originator of CAR/PAR (member of quality assurance department)

(continued)

Activity Description	Output	Responsible
(continued) • Customer complaint report that has been analyzed and approved for remedial action • Findings from internal quality audits, customer audits, or other audits and assessments The originator submits the CAR/PAR to the appropriate management person from the affected area or department and provides two weeks for the receipt of a corrective or preventive action plan, as appropriate. If the plan is not received by the due date, the originator may negotiate a new due date with the concerned management person, or escalate the issue to a higher-level management person in the concerned area or department and within the quality assurance department. The aforementioned escalation mechanism also may be used if the recipient contests the CAR/PAR or otherwise refuses to provide a response.		
2. CAR/PAR response — corrective action plan (CAP) or preventive action plan (PAP) The responsible management person responds to the CAR/PAR by providing the originator with the root causes and proposed corrective or preventive actions in the appropriate form [1, 2].	Root causes and proposed corrective/ preventive actions	Recipient of the CAR/PAR
3. CAP/PAP review The originator reviews the response for agreement with the identified root causes and proposed corrective/preventive actions.		Originator of CAR/PAR

(continued)

Activity Description	Output	Responsible
(continued)		
4. CAP/PAP approved? The originator decides on approval or rejection of the CAP/PAP. If the CAP/PAP is rejected, the originator provides review comments to the responsible management person and a new due date for submission of the revised CAP/PAP. If the CAP/PAP is approved, the process proceeds to step 6.	Decision on approval or rejection of CAP/PAP	Originator of CAR/PAR
5. CAP/PAP revised The responsible management person revises the root causes and/or the proposed corrective/preventive actions as per review comments received from the originator and resubmits them to the originator. (See step 4.)	Revised CAP/PAP	Recipient of the CAR/PAR
6. CAP/PAP implementation Once the CAP/PAP is approved, the management person of the affected area oversees implementation of the CAP/PAP by the due date for completion of the CAP/PAP. The management person is responsible for securing the resources and taking all actions that are necessary for an adequate implementation of the CAP/PAP. Failure of the management person to complete the actions by the agreed-upon due date or a request for additional time will be suitably addressed by the originator of the CAR/PAR; this includes escalation to higher-level management as described earlier in this procedure.	Corrective/preventive actions (implemented)	Recipient of the CAR/PAR
7. Verification of implementation The originator of the CAR/PAR verifies the adequacy and effectiveness of the implemented actions against the approved corrective/preventive action plan.		Originator of CAR/PAR

(continued)

Activity Description	Output	Responsible
(continued)		
8. Implementation acceptable? If the implementation is deemed acceptable, the process proceeds to step 10; if not, additional rework or implementation is requested. This might cause the approved corrective/preventive action plan to be modified with mutual agreement of both parties.	Decision on acceptability of implemented actions	Originator of CAR/PAR
9. Revise implementation. The management person from the concerned area oversees the rework or additional implementation as requested by the originator of the CAR/PAR. Disagreement on additional actions at this stage may result in escalation to higher-level management as described earlier in this procedure.	Rework or additional implementation of corrective/preventive actions	Recipient of the CAR/PAR
10. CAR/PAR closure The originator updates the CAR/PAR by recording the follow-up performed and noting the closure of the CAR/PAR. The originator then submits the closed CAR/PAR for storage in the QMS documentation repository.	Closed CAR/PAR	Originator of CAR/PAR

Section 6: Exit Criteria

- CAR/PAR closed after verification.
- Closed CAR/PAR stored in the QMS documentation repository.

Section 7: Outputs

Output	Receiver
Closed CAR/PAR	QMS documentation repository

Section 8: Process Metrics

- Number of major CARs and PARs opened against each department in each quarter
- Number of minor CARs and PARs opened against each department in each quarter

Section 9: Related Documents

[1] Corrective action request form (Document number: QUA-FR-QMS-0031)
[2] Preventive action request form (Document number: QUA-FR-QMS-0032)
[3] Product defect management procedure (Document number: QUA-PR-QMS-0033)

Section 10: Document Change History

Document Version	*Date*	*Change Performed By*	*Summary of Change*
1.0	05-15-04	Hersh Rahul Nanda	First approved version

APPENDIX C.6: PROJECT POSTMORTEM REVIEW (LESSONS LEARNED) PROCEDURE

Section 1: Introduction

- Purpose
  The purpose of this procedure is to describe the process for handling of lessons learned identified during a project postmortem review.
- Scope
  This procedure describes the complete life cycle for the identification, review, prioritization, implementation, and disposition of lessons learned identified during a project postmortem review.
- Applicability
  This procedure applies only to handling of lessons learned; it does not address handling of other types of situations requiring corrective or preventive action. (These are addressed in the relevant procedures.) Examples of such situations include:
  - Defects identified during verification and validation activities
  - Customer calls (customer complaint, customer-reported problem, or customer suggestion)
  - Corrective and preventive actions necessary to address quality audit findings
  - Corrective and preventive actions necessary during a product development project to address known problems, or to address known and emerging risks
  - Corrective and preventive actions necessary to address improvement suggestions made by employees
- Terminology
  Lesson learned: Lessons learned include opportunities for improvements — problems or deficiencies encountered during a project — and positive experiences (experiences that had an overall positive influence on the outcome of a project).

Section 2: Process Flowchart

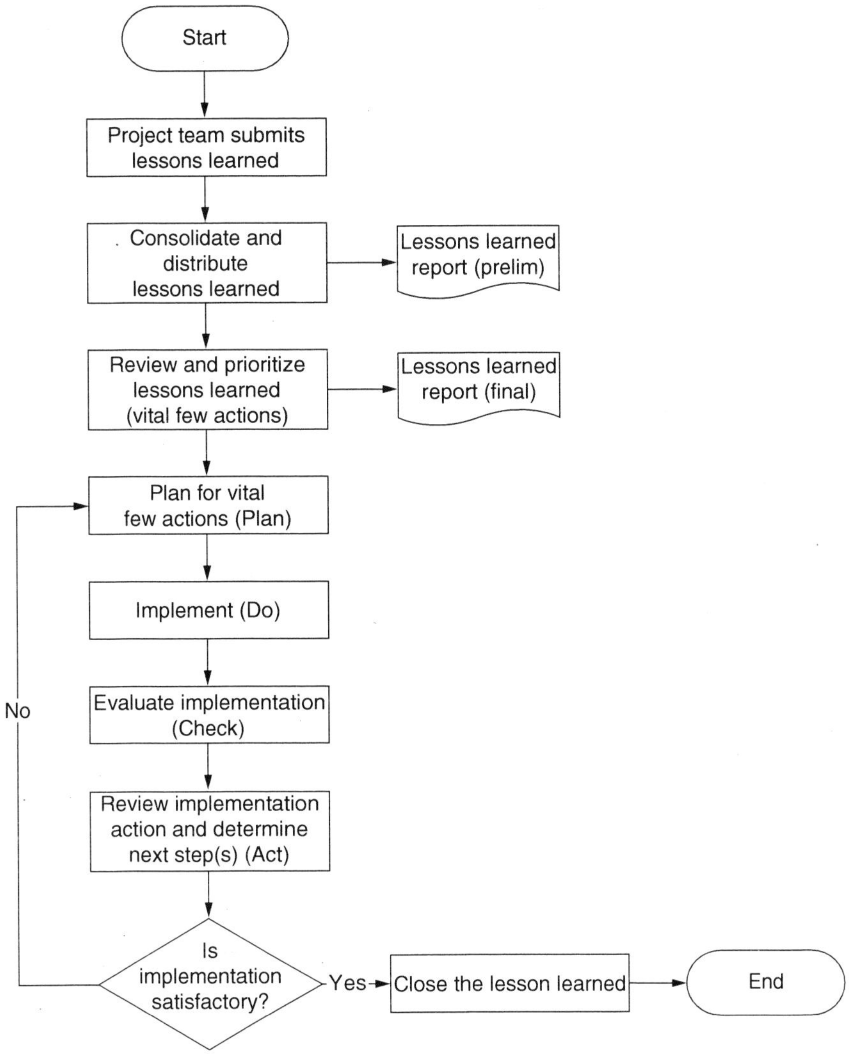

Figure 33 Flowchart for Processing of Lessons Learned

Section 3: Entry Criteria

Completion of a product development project

Section 4: Inputs

Input	*Provider*
Lesson learned (opportunity for improvement or positive experience)	Project participant

Section 5: Activity Descriptions

Activity Description	*Output*	*Responsible*
1. **Project team submits lessons learned** After the completion of a product development project, the quality assurance department requests lessons learned from management personnel from all functional areas involved in the project. The management personnel are provided a due date for providing their feedback and advised to solicit input for this exercise from their personnel who participated in the project.	List of lessons learned provided by each functional area	Project participants
2. **Consolidate and distribute lessons learned** The quality assurance department reviews the lessons learned input received from all parties to identify and eliminate redundancy and to group same or similar issues. The lessons learned are recorded in a spreadsheet; each unique lesson learned is listed on a separate row. Each lesson learned is tagged with a unique ID (for ease of identification)	Lessons learned report (preliminary)	Quality assurance department

(continued)

Activity Description	Output	Responsible
(continued) and the organizational area or process to which the lesson learned is applicable. The frequency of occurrence[a] of each lesson learned also is noted. The frequency of occurrence of a lesson learned is a factor (although not the only one) used to assess the severity (and thus the implementation priority) of the lesson learned.		
The quality assurance department compiles the list of lessons learned separately as opportunities for improvement and positive experiences, in a formal report that serves as the review material for a lessons learned meeting attended by management personnel from all functional areas.		
3. **Review and prioritize lessons learned (vital few actions [VFAs]).**		
The quality assurance department facilitates a lessons learned meeting to review and prioritize lessons learned. A prioritization of the lessons learned is necessary because often the number of lessons learned far exceeds the number that can be addressed, due to resource constraints, return-on-investment considerations, or other factors. The lessons learned meeting entails the following:	Lessons learned report (final) with list of selected lessons learned, and positive experiences	Lessons learned meeting participants
• Each opportunity for improvement is discussed for validity, impact (on product quality, organizational efficiency, and effectiveness),		

(continued)

Activity Description	Output	Responsible
(continued)		
and severity (based on frequency of occurrence or severity of impact). Each opportunity for improvement is prioritized as high, medium, or low. It also is determined if an improvement opportunity is already being addressed by current or planned improvement efforts. Lessons learned that require immediate attention or that can be implemented relatively easily are selected for implementation and assigned to appropriate management personnel. The person responsible for the lesson learned thereafter is responsible for planning and implementation of needed actions (VFAs). Meeting participants may also suggest corrective or preventive actions for the selected lessons learned.		
• Each positive experience is discussed, and it is determined whether the positive experience was unique to the project or can be replicated in future projects. If the latter, it is determined if a process or procedural change is necessary to ensure that the positive experience is repeated. For such items, an appropriate management person is identified for implementation of needed actions.		

(continued)

Activity Description	Output	Responsible
(continued)		
Note: The corrective and preventive action procedure may be used for initiating and following up on required corrective and preventive actions. (Refer to [1], [2], and [3].)		
4. Plan for VFAs (Plan) The responsible management person prepares the corrective or preventive action plan necessary to address the lesson learned, and provides it to the quality assurance department for review. When appropriate, management personnel from other functional areas may be involved in the review of the corrective/preventive action plan. The plan is revised to the satisfaction of all reviewers (stakeholders) prior to its implementation.	Corrective/preventive action plan for each lesson learned	Management person responsible for the lesson learned
5. Implement (Do) The appropriate management person is responsible for the implementation of the plan after its review and approval by the quality assurance department and other parties, as appropriate. The corrective or preventive action generally entails a process and/or procedural change and its execution at the next opportunity (typically the next product development project) in order to determine the adequacy and effectiveness of the actions taken.		Management person responsible for the lesson learned

(continued)

Activity Description	Output	Responsible
(continued)		
6. **Evaluate implementation (Check)** Implementation results are examined to determine whether or not the implementation meets expected results, and appropriate corrective action is taken, as needed.		Management person responsible for the lesson learned
7. **Review implementation action and determine next steps (Act)** Implementation results are reviewed with the quality assurance department and other stakeholders, as appropriate.	Implementation results	VFA reviewers
8. **Is implementation satisfactory?** If the implementation is satisfactory, the VFA is closed (see step 9); otherwise, additional actions are identified and the process returns to step 4 for replanning.	VFA decision	VFA reviewers
9. **Close the lesson learned** The quality assurance department closes the lesson learned after successful implementation.		Quality assurance department

[a] This refers to the number of times the same (or similar) lesson learned is cited independently by different project participants. For example, if a lesson learned is cited by three different management personnel, the frequency of occurrence of this lesson learned is three.

Section 6: Exit Criteria

Lesson learned (closed) after implementation

Section 7: Outputs

Output	Receiver
Lessons learned reports (preliminary and final)	QMS documentation repository

Section 8: Process Metrics

Number of lessons learned implemented each year in each functional area (ordered by severity)

Section 9: Related Documents

[1] Corrective action request form (Document number: QUA-FR-QMS-0031)
[2] Preventive action request form (Document number: QUA-FR-QMS-0032)
[3] Corrective and preventive action procedure (Document number: QUA-PR-QMS-0030)

Section 10: Document Change History

Document Version	*Date*	*Change Performed By*	*Summary of Change*
1.0	2-01-04	Ellora Adams	First approved version
1.1	6-05-04	Ellora Adams	Included examples of when this procedure is not applicable. Included other minor clarifications and corrections.

APPENDIX C.7: PROCEDURE FOR THE DEPLOYMENT OF CORPORATE QUALITY OBJECTIVES

Section 1: Introduction

- Purpose
  The purpose of this procedure is to describe the process for the deployment of corporate quality objectives into operational plans in the organization.
- Scope
  The scope of this procedure covers the initial identification of high-level corporate quality objectives, their systematic breakdown into quantitative quality objectives for lower levels of the organization, and the execution of improvement projects to achieve the defined objectives.
- Applicability
  This procedure is applicable to management personnel from all functional areas in the organization.
- Terminology
  None

Section 2: Process Flowchart

None

Section 3: Entry Criteria

Request from the QMS steering group to begin the annual corporate quality diagnosis

Section 4: Inputs

Input	*Provider*
QMS metrics report for previous calendar year	Quality assurance department

Section 5: Activity Descriptions

Activity Description	Output	Responsible
1. **Define high-level quality objectives**		
At the beginning of each calendar year, senior and middle management conduct an annual corporate quality diagnosis to examine quantitative and qualitative information regarding organizational process performance and to formulate annual quality objectives. In this context, annual quality objectives are regarded as short-term quality objectives. Management personnel also review organizational performance against existing long-term quality objectives (two to three years), establish new long-term quality objectives as needed, and revise existing long-term quality objectives when necessary. The goal at this "macro" level is to define the strategic vision of top-level management with regards to quality and organizational performance to achieve a competitive advantage and organizational excellence.	High-level quality objectives	Senior and middle management
2. **Break down high-level quality objectives into lower-level quality objectives**		
Once high-level corporate quality objectives have been established, middle and lower management incrementally elaborate and decompose them into SMART goals (and supporting improvement actions) for lower levels of the	Lower-level quality objectives	Middle and lower management

(continued)

Activity Description	Output	Responsible
(continued)		
organization. Each improvement action is assigned an appropriate management person who has overall responsibility for the implementation of the improvement action.		
For example, at the highest level, a corporate quality objective might be defined as "Improve customer satisfaction survey rating in the following year by an average of 5% for all surveyed items." At the next lower level, a quality objective may be defined as "Improve customer satisfaction rating for 'quality of customer service' survey item by 10% in the following year." The supporting improvement action may state "Provide advanced five-day product-specific training for all customer support personnel," and "Record the response time for all customer calls. Review response time data with department management on a biweekly basis to investigate and correct causes of delay beyond stipulated response times (specified in company quality procedures)."		
3. **Vertical and horizontal review of defined objectives**		
Once lower-level quality objectives have been defined, consistency between planned improvement actions across units (at the same level or at different levels) must be ensured. The mechanism for reporting progress from lower to senior management		Middle and lower management

(continued)

Activity Description	Output	Responsible
(continued) regarding deployment of the quality objectives also should be defined. The main activities in this step are: • *Horizontal handshake:* During this activity, the manager responsible for the higher-level quality objective meets with all managers with associated lower-level objectives and proposed improvement actions to review the underlying objectives and proposed actions for cross-functional consistency. During this step, gaps, mismatches, and other inconsistencies are identified. • *Vertical handshake:* Once the cross-functional discrepancies have been resolved, the proposed improvement actions are reviewed and negotiated with the manager responsible for the higher-level improvement objective. This ensures vertical consistency and the ability to meet the higher-level improvement objective. • *Define review and reporting process:* This involves the definition of the reporting mechanism to be used once the implementation gets under way.		

(continued)

Activity Description	Output	Responsible
(continued)		
4. Prepare improvement project plan		
Each improvement action (or related improvement actions) spawns one or more improvement projects (usually one) to address the action and take it to a successful closure. The responsible management person prepares an improvement project plan that specifies the complete list of planned subactions and evaluation mechanisms to ensure the successful deployment of the related improvement action. The improvement project plan contains all pertinent information related to a specific project, such as: • Background information on the project and analysis of causes that led to the improvement project • Specific goal of the improvement project • Major actions planned to bring the improvement action to a successful closure • Measurements to be used to assess the success of the project • Schedule for the various phases of the project • Resource allocation • Mechanism for reporting progress	Improvement project plan	Responsible management person
5. Execute improvement project		
This is perhaps the most important activity of the process, and involves the execution of each of the	Improvement project outcome	Responsible management person

(continued)

Activity Description	*Output*	*Responsible*
(continued) improvement projects as per the defined plan. During implementation, the responsible management person reports progress to senior management as per the identified reporting mechanism. Results of the improvement project are compared with the expected results and a decision as to whether or not additional tasks are necessary to successfully conclude the project is made. In case of rework, additional tasks deemed necessary are identified and executed, and results are re-evaluated to determine the outcome of the improvement project. If the improvement objective cannot be met upon conclusion of the project, a reexamination of the feasibility of the improvement objective might be necessary, as might identification of additional improvement projects necessary to meet the objective.		

Section 6: Exit Criteria

Conclusion of the improvement project

Section 7: Outputs

Output	*Receiver*
Improvement project results	QMS steering group

Section 8: Process Metrics

None

Section 9: Related Documents

None

Section 10: Document Change History

Document Version	*Date*	*Change Performed By*	*Summary of Change*
1.0	06-10-04	Ronnie Small	First approved version

Appendix D

SAMPLE FORMS AND TEMPLATES

APPENDIX D.1: PROCEDURE TEMPLATE

Section 1: Introduction

- Purpose
  <Briefly describe the purpose of the process that this procedure describes.>
- Scope
  <Identify the scope of this process.

  This section also should be used to identify activities that are beyond the scope of this procedure (but that could be assumed to be within its scope without this important caveat).>
- Applicability
  <Identify functional areas (departments) in the organization to which this procedure applies. This information subsequently also can be used to identify relevant personnel who need to be notified in case of changes to this document, and who need to be provided training on this process.

  This section also should be used to identify situations or areas to which this procedure does not apply (but that could be assumed to be applicable to without this important caveat).>

- Terminology
  <Only define terms that are specific to this document, or not otherwise defined in the organization's formally documented QMS Glossary.>*

Section 2: Process Flowchart

<For documentation of procedures, it is highly desirable to represent the process map by means of a flowchart that uses standardized notation. Flowcharts help readers quickly comprehend a process by visually understanding the process flow. The flowchart should include all major activities, or steps, that are included in the process. Each flowchart activity must be supported by a corresponding written description of the activity in Section 5. Figure 34 presents an example of a flowchart notation that is available in popular word-processing software.

A flowchart may be omitted if the process flow is sequential, and providing a flowchart does not enhance readability and understanding.>

Section 3: Entry Criteria

<List all inputs and approvals that are necessary for the start of this process.>

Section 4: Inputs

<List all inputs to this procedure and identify the associated suppliers — i.e., the process, or procedure, or role, or department providing the input.>

---

* It is recommended that, as part of the QMS documentation, a QMS glossary be formally documented and approved for organization-wide use. This will help facilitate a common understanding and consistent use of terminology in any communication and documentation pertaining to the QMS. Such a document is especially necessary in very large organizations, or organizations with multiple locations.

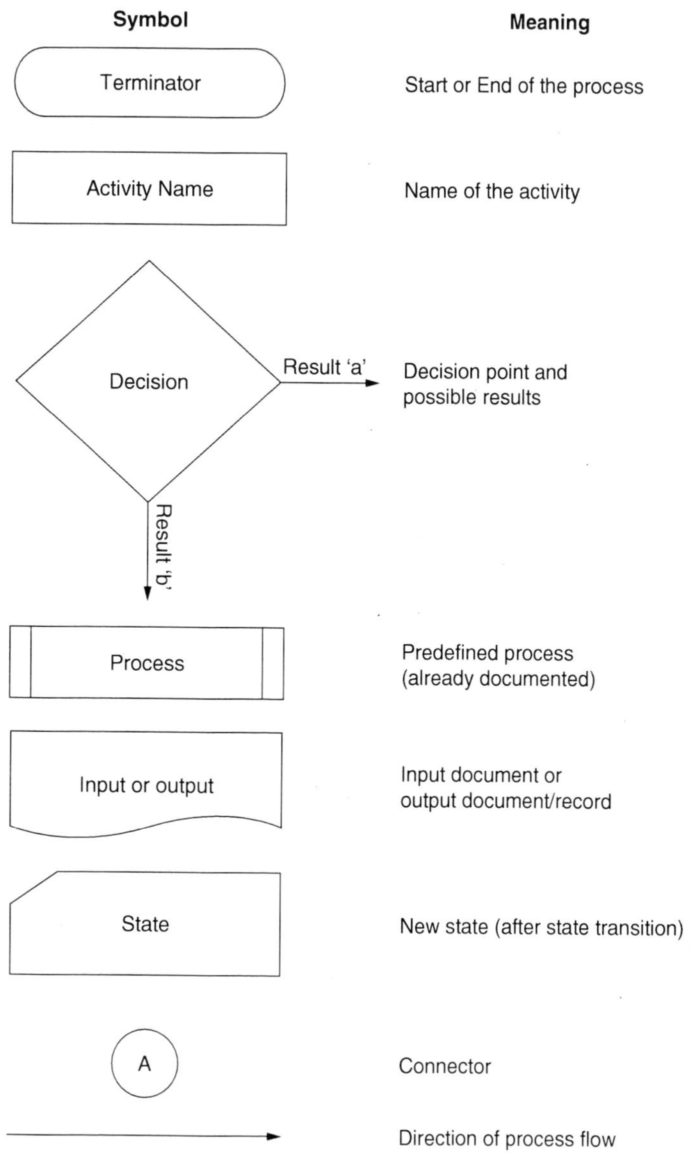

Figure 34 Flowchart Notations

Section 5: Activity Descriptions

Activity Description	*Output*	*Responsible*
<Describe each major activity in a separate row; ensure that the list of activity names matches those in the flowchart>	*<List outputs for each activity>*	*<Identify by role or department name who is responsible for performing the activity>*

Section 6: Exit Criteria

<List all outputs and approvals that are necessary for the end of this process.>

Section 7: Outputs

<List all outputs of this procedure and identify the associated customers — i.e., the process, or procedure, or role, or department receiving the output.>

Section 8: Process Metrics

<If appropriate, list process metrics that have been established to measure, monitor, and continually improve this process.>

Section 9: Related Documents

<List other documents relevant to this procedure. These may include templates, guidelines, checklists, and the like.>

Section 10: Document Change History

Document Version	*Date*	*Change Performed By*	*Summary of Change*
<Identify each document version>	*<Identify the date each document version was created>*	*<Identify who performed the change for each document version>*	*<Provide a brief summary of changes for each document version>*

APPENDIX D.2: CORRECTIVE ACTION REQUEST (CAR) FORM*

<table>
<tr><td colspan="2">CAR #:</td><td colspan="2">Severity: Major ☐ Minor ☐ Observation ☐</td></tr>
<tr><td colspan="4" align="center">Nonconformance Summary</td></tr>
<tr><td>Department/Process Audited:</td><td colspan="2">Auditor:</td><td>Date:</td></tr>
<tr><td>Auditee Manager:</td><td colspan="3"></td></tr>
<tr><td colspan="4" align="center">Reference</td></tr>
<tr><td colspan="4">Quality management system standard: <<i>Identify the clause/section in the applicable industry QMS standard against which the nonconformance is noted.</i>></td></tr>
<tr><td colspan="4">Quality Management System document and version: <<i>Identify the QMS document against which the nonconformance is noted. If the nonconformance is against a project plan document, identify it here.</i>></td></tr>
<tr><td colspan="4" align="center">Nonconformance Description</td></tr>
<tr><td colspan="4">Description:</td></tr>
<tr><td colspan="4" align="center">Corrective Action Plan</td></tr>
<tr><td colspan="4">Root Cause:</td></tr>
<tr><td colspan="4">Planned Corrective Action:</td></tr>
<tr><td colspan="3">Signature of Auditee's Supervisor:</td><td>Estimated date of completion:</td></tr>
</table>

(continued)

* Reproduced with permission from Nanda, Vivek (Vic), *ISO 9001:2000 - Achieving Compliance and Continuous Improvement in Software Development Companies*, ASQ Quality Press, 2003.

Appendix D.2 (continued)

Approved by Auditor: Yes ☐ No ☐	Actual date of completion:
Corrective Action Follow-up	
Date of follow-up:	
Follow-up comments:	
Additional follow-up required? Yes ☐ No ☐	Date of additional follow-up (if applicable):
CAR closed? Yes ☐ No ☐	Auditor:

APPENDIX D.3: DOCUMENT CHANGE REQUEST (DCR) FORM

DCR #:	Severity: Major ☐ Minor ☐	Date Submitted:
	Document Information	
Document Name:	Document Number:	Doc. Version:
Other documents affected: *<List document name, document number, and document version of other documents affected by the change being requested to this document. If this DCR is approved, separate DCRs for the affected documents should be opened (unless the changes to other affected documents are described and approved under this DCR).>*		
	Submitter Information	
Name:	Department:	
	Change Request Details	
Description of change: *<Provide a description of the change.>*		
Reason for change: *<Provide an explanation for why the change is being requested.>*		
	Decision	
Approved: ☐ Rejected: ☐		
	Approved DCRs	
DCR Approved by:		
☐ Quality Assurance	Name of approver:	Date approved:
☐ Product Engineering	Name of approver:	Date approved:
☐ Product Marketing	Name of approver:	Date approved:
☐ Production	Name of approver:	Date approved:

(continued)

Appendix D.3 (continued)

<table>
<tr><td colspan="3"><Add additional rows to list names of other departments that participate in the PMC and might need to approve a document change request. For a specific DCR, those that apply should be checked, and the name of the approver and the date of the approval should be recorded.></td></tr>
<tr><td colspan="3">Rejected DCRs</td></tr>
<tr><td colspan="3">Reason for rejection:</td></tr>
<tr><td colspan="3">DCR Assigned to (Approved DCRs only)</td></tr>
<tr><td rowspan="2">Name: <The DCR may be assigned to the process owner or document author.></td><td rowspan="2">Department:</td><td>Due date:</td></tr>
<tr><td>Actual Completion date:</td></tr>
</table>

APPENDIX D.4: TEMPLATE FOR MINUTES OF MEETING FOR *<PROJECT NAME>*

Distribution List:

<List names of personnel who are to receive these minutes of meetings.>

Meeting Details:

Last Meeting Held: *<Provide the date, time, and location of the most recent meeting.>*

Meeting Participants:

<For each participant, record the name and department of the participant in the table below. Add as many rows as needed.>

Name	*Department*

Next Meeting:

<Provide the date, time, and location for the next meeting.>

Meeting Agenda:

<List the agenda items for the meeting. List each agenda item separately on each line.>

1. *<First agenda item>*
2. *<Second agenda item>*
3. *<Third agenda item>*
4. *<and so on…>*

Meeting Minutes:

<For each agenda item, provide minutes in separate paragraphs as shown below.>

1. *<First agenda item>*
   <Provide a brief summary of what was discussed, decisions pertaining to this agenda item, and numbered references to related action items in the action items table in the next section.>
2. *<Second agenda item>*
   <Provide similar information as for the first agenda item. Repeat this process until all agenda items are addressed.>

Action Items:

Action Item	*Description*	*Responsible*	*Notes*	*Due Date; Date Closed*
1.	*<Briefly summarize the action item>*	*<Identify responsible person>*	*<Record pertinent information, such as related issues, progress made, and so on>*	*<Record due date and date of closure for this action>*
2.	...			

APPENDIX D.5: PROJECT MILESTONE REVIEW REPORT FORM

Project Name:	*<Record the name of the project.>*
Milestone (MS):	*<Identify the formal project milestone for which this milestone assessment is being performed.>*
Planned MS date:	*<Record the planned date of this milestone assessment (from the approved project plan).>*
Actual MS date:	*<Record the actual date of this milestone assessment.>*

MS Assessment Result (check one): Pass* _____ Fail** ______

Checklists:

MS Approval Checklist:

- Verify that all approved documents have been submitted for storage in the approved documents repository.
- Verify that all unapproved and unavailable deliverables have been included for follow-up in the next project review meeting, next milestone review meeting, and/or the project risk management plan (as appropriate).
- Identify which (if any) project tasks from subsequent phases have been put on hold due to unapproved or unavailable deliverables.

Failed MS Checklist:

- Reasons for failure: *<State why this milestone assessment has failed.>*
- Verify that all unapproved and unavailable deliverables have been included for follow-up in the next project review meeting, next milestone review meeting, and/or the risk management plan (as appropriate).
- Identify which (if any) project tasks from the subsequent phases are being allowed to progress further.
- Next date for MS reassessment: *<Identify the date when this milestone will be reassessed.>*

Action Items:

* A milestone assessment passes if all required deliverables are available and approved or, in the case of unavailable or unapproved deliverables, the deviations are approved by the milestone assessment team (due to minimal risk to future project activities).

** The milestone assessment fails if the milestone assessment team deems a re-assessment necessary due to one or more unapproved deviations.

MS Deliverables	*Doc. Owner*	*Deliverable Available? [Yes/No]*	*Doc. Number and Version*	*Deliverable Approved? [Yes/No]*	*Deviation Approved? [Yes/No]*
<List each project document or deliverable that was due at this milestone; list each on a separate line>	*<Record the name of the department that owns the deliverable>*	*<Is the required deliverable available?>*	*<If the deliverable is available, record its number and version>*	*<Is the deliverable approved?>*	*<If the deliverable is not available or approved, is this deviation approved[a] by those conducting the milestone assessment?>*

[a] Relevant management personnel from departments participating in a project typically perform a milestone assessment. All deviations must be approved for a project to be allowed to progress to the next phase. In certain cases where a deliverable (or set of deliverables) is not deemed to be on the critical path of a project (that is, nonavailability of the deliverables does not jeopardize further progress), deviation may be granted by consensus of the meeting participants.

Action Item	*Description*	*Responsible*	*Notes*	*Due Date; Date Closed*
1.	*<Briefly summarize the action item>*	*<Identify responsible person>*	*<Record pertinent information, such as related issues, progress made, and so on>*	*<Record due date and date of closure for this action>*

APPENDIX D.6: REVIEW RECORD FORM

Review Information:

Project/product name:	
Name of artifact being reviewed:	
Number and version of artifact:	
Review date:	
Type of review:	Walkthrough ☐ Inspection ☐
Review/rereview:	Review ☐ Rereview ☐
Meeting participants: Author (Name/Department): Moderator (Name/Department): Recorder (Name/Department): Reviewers (Name/Department):	
Total major remarks:	
Total minor remarks:	
Total man–hours meeting time:	
Rework due date:	
Review decision (choose one): Approved (subject to rework) ☐ Failed (rework & rereview required) ☐	

Review Remarks:

#	*Location (Page/ Section)*	*Severity (Major[a]/Minor[b]/ Out of Scope)*	*Action Item*	*Responsible*	*Action Taken*
					<Record what action was taken to address this action item>

[a] A major remark pertains to an error, omission, or required major clarification. [b] A minor remark pertains to a typographical error, grammatical error, spelling error, reviewer suggestion, or required minor clarification.

Rework Information:

Rework completed:	Yes ☐ No ☐
Comments on exceptions (if any):	
Moderator's signoff:	

Appendix E

EXAMPLE OF AN AUDIT CHECKLIST

Area(s) audited:	*<Specify the area and/or department for which this audit checklist has been created.>*
Date of audit:	*<Specify the date of the audit.>*
Auditor:	*<Specify the name of the auditor(s).>*
Auditee:	*<Specify the name of the auditee(s).>*
Auditee Supervisor:	*<Specify the name of the auditee supervisor.>*

Note: This checklist contains a sample list of questions that may be asked in an audit. Not all the questions in this checklist must be asked in an audit. Further, the auditor may ask additional questions that are not included in this checklist.

		Finding (check one)				
ID	*Audit Item*	*Major Remark*	*Minor Remark*	*Observation*	*Reference (Objective Evidence)*	*Statement of Finding or Auditor's Notes*
1.	*<Specify the first question that might be asked during the audit>*				*<Specify the objective (verifiable) evidence that was examined for this audit item — for example, document, quality record, equipment, etc.>*	*<In case of a finding, this space may include the following information:*[a] **Statement of finding:** *<Auditor specifies the statement of finding>* **Root causes:** *<Auditee supervisor specifies the root causes for the observed condition>* **Corrective/Preventive Action Plan:** *<Auditee supervisor provides a corrective and/or preventive action plan that effectively addresses the root causes for the observed condition>*

Corrective/Preventive Action Follow-Up:
<Auditor includes notes from verification of the implemented actions and explicitly states if the finding is closed>

<If no finding is reported for this audit item, this space may be used by the auditor to record pertinent audit notes, such as auditor's suggestions, to acknowledge recent improvements, etc.>

2. And so on ...

[a] Alternatively, this information may be included in a separate form for documenting audit findings and requesting corrective action, such as the Corrective Action Request Form in Appendix D.2.

Date audit report closed: *<Date of closure of audit report after satisfactory implementation of all corrective and preventive actions.>*

Audit report closed by: *<Name of auditors closing the report.*>*

* The audit may be closed by the original auditor, or by another qualified auditor who performed the audit follow-up.

Appendix F

SAMPLE TRAINING MATERIAL

APPENDIX F.1: EXAMPLE OF A TRAINING COURSE CATALOGUE

This training catalogue contains three levels of training courses:

- **QMS 100 courses:** These courses are foundation courses that cover fundamental topics pertaining to the QMS. QMS 100 courses are an essential prerequisite for QMS 200- and 300-level courses.
- **QMS 200 courses:** These courses provide information necessary to execute work processes in accordance with applicable requirements.
- **QMS 300 courses:** These courses are hands-on workshops to impart specialized employee skills necessary for executing specific tasks.

QMS 100 Courses — Foundation Courses

Description: QMS 100 courses serve as the necessary foundation for an employee's understanding of the QMS. This series includes a course that provides an overview of the QMS, and additional courses that provide in-depth information on key elements of the QMS.

Applicability: QMS 100 courses are mandatory for all employees of the organization.

QMS 100: Quality Management System Overview	This course provides an overview of the QMS and includes a demonstration of the electronic QMS documentation repository.

(continued)

(continued)

QMS 110: Document Management and Control Process	This course provides an explanation of the concepts fundamental to document management and control, including controlled documents, document numbering and versioning, document approval, document changes (including emergency changes), control of external documents, and storage of approved documents.
QMS 120: Understanding Procedures and Document Templates	This course examines in detail the structure and expected content of the various sections in a procedure and helps facilitate employee understanding of procedures. This course also provides information necessary to understand and correctly use document templates.

QMS 200 Courses — Work Processes

Description: QMS 200 courses provide employees with information on the management processes, technical processes, and support processes relevant to their jobs.

Applicability: Refer to the QMS training tracks for information on which courses are applicable to which employees.

QMS 205: Planning, tracking, and Controlling Product Development Activities	This course covers topics relevant to project management. Specifically, it describes the process for planning for product development, and techniques for tracking and monitoring progress and for controlling product development activities. The participants of this course will manage a mock project and learn project management techniques by direct application.
QMS 210: Supplier Excellence — Supplier Selection, Management, and Control	This course describes a systematic process for supplier selection, and techniques for effectively monitoring and controlling supplier performance for continued supplier excellence.
QMS 215: Defining Product Requirements	This course describes techniques for eliciting and defining high-quality requirements that are complete, correct, unambiguous, and testable. Participants will learn requirements definition with exercises that mimic real-life scenarios.
QMS 220: Product Design and Implementation	This course describes the process of product design and implementation.

(continued)

(continued)

QMS 225: Product Verification and Validation	This course describes various tasks that may be performed during the product development process to verify product requirements, design, and implementation, and to validate the end product against product requirements.
QMS 230: Contract Reviews	The course describes the process of contracts definition and reviews. The course also describes the process for executing an amendment to an approved contract.
QMS 235: Interfacing with Customers — Tools and Techniques for Effective Communication	This course describes different vehicles for communicating with customers, and it describes the purpose of each type of communication vehicle. The course also provides guidelines on how to make the overall experience a positive one for the customer.
QMS 240: Employee Competency Development	The course gives in-depth guidance on establishing job competency requirements, assessing employee competency and needs, enhancing employee competency, and assessing effectiveness of actions taken.

QMS 300 Courses — Workshops

Description: Workshops provide essential skills for performing specific functions required for the QMS.

Applicability: These workshops are mandatory for employees involved in these functions.

QMS 310: Procedure Workshop	This workshop provides a hands-on introduction to writing procedures with the aid of real-life examples.
QMS 320: Moderator Workshop	This workshop provides a hands-on introduction to moderating formal reviews. Participants learn by moderating mock reviews of documents, and by observing effective and ineffective techniques for moderating reviews.
QMS 330: Audit Preparation Workshop	This workshop provides an introduction to the audit process, with emphasis on what to expect during external audits, including guidance on desired auditee behavior, auditing techniques (employed by the auditor), and types of questions to expect during an audit interview.

APPENDIX F.2: EXAMPLE OF A COURSE DESCRIPTION

Course Name: QMS 100: Quality Management System Overview

Course Objective	This course provides an overview of the QMS and includes a demonstration of the electronic QMS documentation repository.
Audience	This course is mandatory for all employees of the organization. *Prerequisite:* Nil
Presentation Logistics	*Number of slides:* 33 *Course handouts:* • Copy of Quality Manual • Product development process map • Product development plan template • Project quality plan template • QMS Quick Reference Card *Time Required:* 2 hours (without break) *Class size:* 5 to 30 *Equipment required:* • Laptop • Projector and screen • Pointer • Flipchart and marker *Qualified Instructors:* Bill Rippone and Kathy Lee
Course Agenda	QMS basic concepts: • Definitions • Benefits of a QMS • Functional view vs. Process view • QMS documentation hierarchy • Types of QMS documentation QMS implementation process Quality practices applicable to different organizational areas (overview only): • Management processes • Technical processes • Support processes Q & A
Course Follow-up	A classroom quiz will be administered at the end of the course. (A minimum score of 75% is required for course credit.)

APPENDIX F.3: EXAMPLE OF QMS TRAINING TRACKS

Note: This table illustrates the applicability of QMS training courses to various types of roles in an organization, each of which constitutes a separate audience track. The table is for illustration purposes only and is not intended to depict all roles that typically exist in a product development company.

Training Tracks ⇒ *QMS Training Courses ⇓*	*Track 1* *Project Managers*	*Track 2* *Buyers*	*Track 3* *Quality Assurance Personnel*	*Track 4* *Requirements Engineers*	*Track 5* *Product Developers*	*Track 6* *Product Testers*	*Track 7* *Contract Administrators*	*Track 8* *Human Resource Specialists*	*Track 9* *Customer Service Staff*
QMS 100 Training Courses									
QMS 100: Quality Management System Overview	X	X	X	X	X	X	X	X	X
QMS 110: Document Management and Control Process	X	X	X	X	X	X	X	X	X
QMS 120: Understanding Procedure Descriptions and Document Templates	X	X	X	X	X	X	X	X	X
QMS 200 Training Courses									
QMS 205: Planning, Tracking, and Controlling Product Development Activities	X								
QMS 210: Supplier Excellence — Supplier Selection, Management, and Control		X	X						

QMS 215: Defining Product Requirements				X					
QMS 220: Product Design and Implementation					X				
QMS 225: Product Verification And Validation					X	X			
QMS 230: Contract Reviews							X		
QMS 235: Interfacing with Customers — Tools and Techniques for Effective Communication									X
QMS 240: Employee Competency Development								X	
QMS 300 Training Courses									
QMS 310: Procedure Workshop	X	X	X	X	X	X	X	X	X
QMS 320: Moderator Workshop	X		X	X	X	X			X
QMS 330: Audit Preparation Workshop	X	X		X	X	X	X	X	X

APPENDIX F.4: EXAMPLE OF QMS OVERVIEW TRAINING MATERIAL

QMS 100
Quality Management System Overview

Instructor: <*Name of instructor*>

Date: <*Date of presentation*>

Course Agenda

- QMS basic concepts
  - – Definitions
  - – Benefits of a QMS
  - – Functional view vs. Process view
  - – QMS documentation hierarchy
  - – Types of QMS documentation
- QMS implementation process
- Quality practices applicable to different organizational areas (overview only)
  - – Management processes
  - – Technical processes
  - – Support processes
- Q & A

Definitions

- Quality
- Quality planning
- Quality control
- Quality improvement
- Quality management
- Quality management system

Benefits of a QMS

- Improvement in operational efficiency
- Improvement in organizational effectiveness
- Higher-quality products and services
- Process focus
- Continuous improvement
- Customer satisfaction
- Higher customer loyalty and retention
- Competitive advantage
- Reduced waste
- Reduced dependence on heroes
- Reduced organizational vulnerability

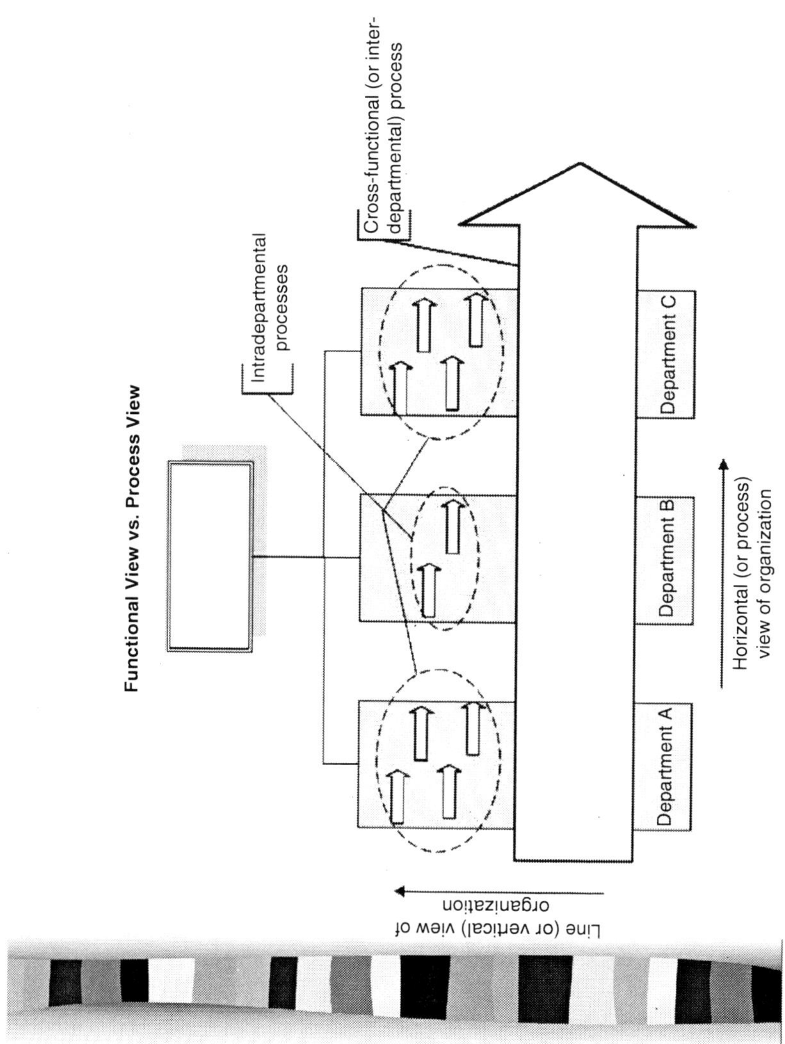
Functional View vs. Process View
Intradepartmental processes
Cross-functional (or inter-departmental) process
Department A
Department B
Department C
Horizontal (or process) view of organization
Line (or vertical) view of organization

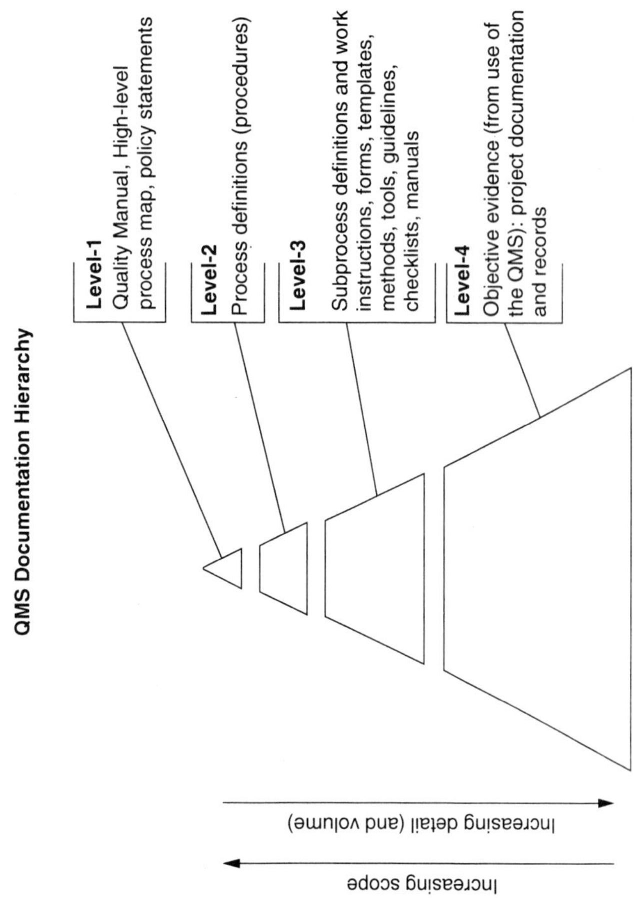
QMS Documentation Hierarchy
Level-1
Quality Manual, High-level process map, policy statements
Level-2
Process definitions (procedures)
Level-3
Subprocess definitions and work instructions, forms, templates, methods, tools, guidelines, checklists, manuals
Level-4
Objective evidence (from use of the QMS): project documentation and records
Increasing detail (and volume)
Increasing scope

Types of QMS Documentation

- Quality Manual
- Procedures
- Work Instructions
- Forms and Templates
- Project Documents and Records
- Checklists, Guideline documents, & workmanship standards

QMS Implementation Process

- Phase I: Plan
  - QMS implementation planning
  - QMS documentation planning
- Phase II: Define
  - Defining organizational processes
  - Quality practices
- Phase III: Refine
  - QMS refinement
- Phase IV: Deploy
  - QMS deployment
- Phase V: Improve
  - Continuous improvement

Organization Chart

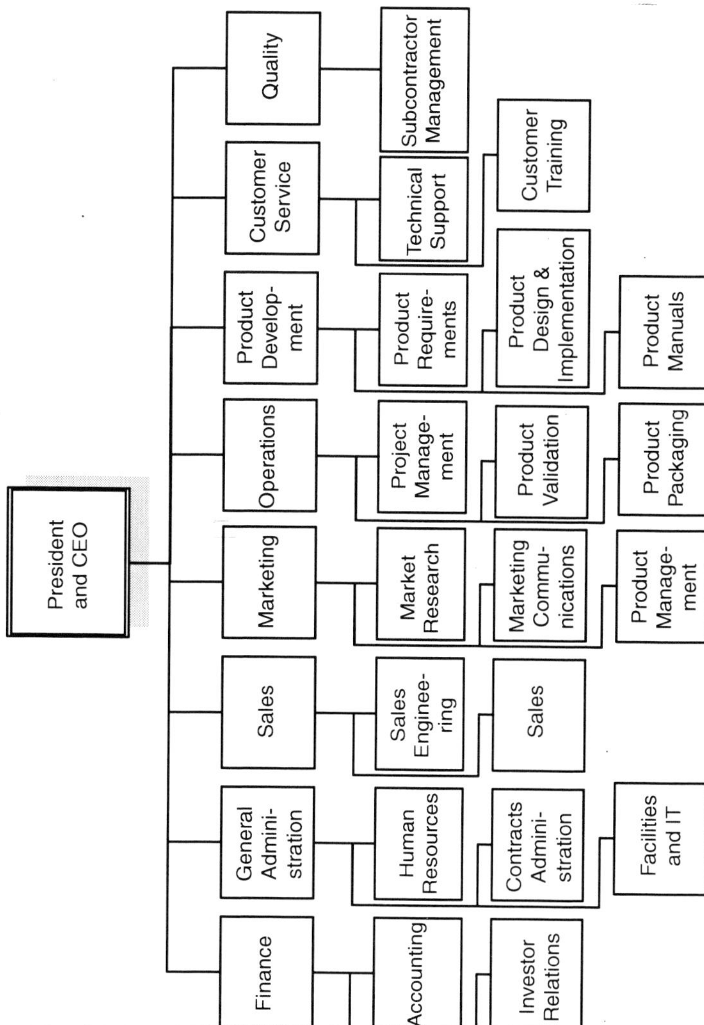

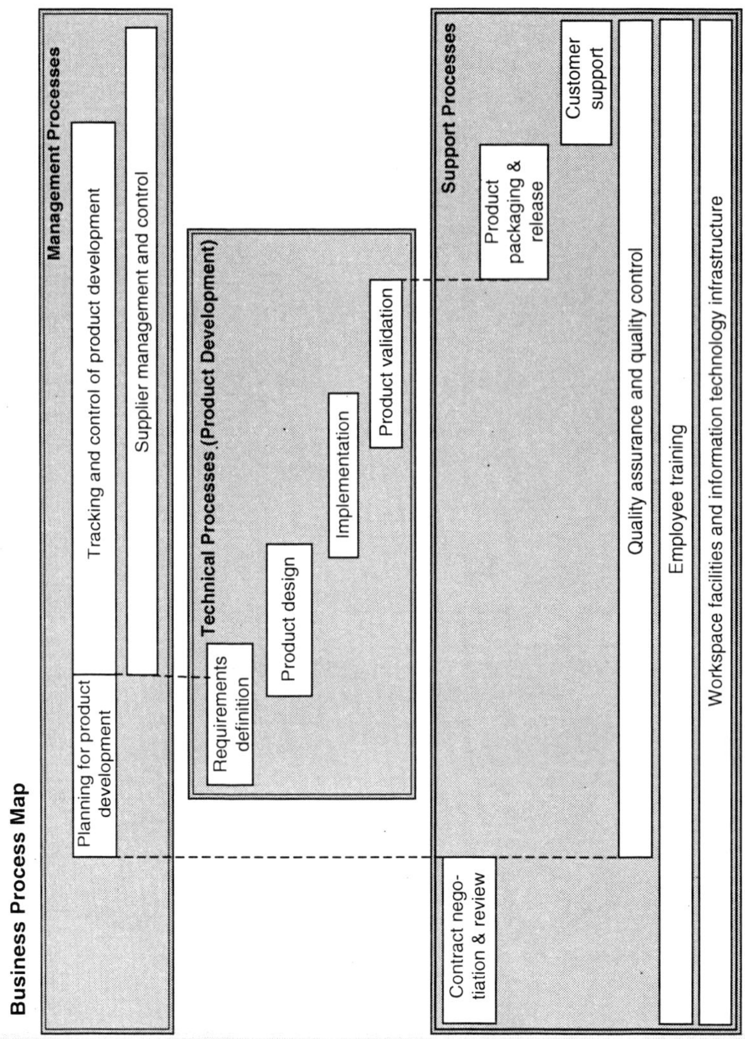
Business Process Map
Management Processes
Planning for product development
Tracking and control of product development
Supplier management and control
Technical Processes (Product Development)
Requirements definition
Product design
Implementation
Product validation
Support Processes
Contract negotiation & review
Product packaging & release
Customer support
Quality assurance and quality control
Employee training
Workspace facilities and information technology infrastructure

Management Processes

- Planning for product development
- Tracking and control of product development
- Supplier management and control

Management Involvement

- Senior management has overall responsibility for QMS
- A management body has been established to exercise management oversight over the QMS
- Bill Smart is the quality management representative
- QMS management reviews are held bi-monthly
- Sample items covered at management reviews:
  - –Short-term and long-term quality objectives
  - – Status of corrective/preventive actions
  - – Status of capability improvement initiatives
  - – Measurement results and trends
  - – Results of internal and external audits and assessments
  - – Customer satisfaction data
  - – Supplier performance
- Records of management reviews of QMS are maintained

Planning for Product Development

- Process for product development planning has been documented
- Product development plan template has been established (see handout)
- Project quality plan template has been established (see handout)
- Standardized QMS documentation has been created to guide execution of product development and support activities (see handout of Quality Manual)

Tracking and Control of Product Development

- Project execution is closely monitored by project manager:
  - Separate weekly project review meeting held for each project
  - Project action items are recorded in the "Project action items spreadsheet"
  - Project risks and associated mitigation and contingency plans are monitored with project risk management tool
  - Formal milestone review meetings held at major project milestones
- Line manager tracks resource utilization formally by managing employee resource pools
- Project manager maintains product development plan and revises it to reflect replanning (as needed)

Supplier management and control

- Suppliers are selected, monitored, and controlled in accordance with supplier management and control procedure:
  - Procedure applies to items (products or parts) purchased for use in product development
  - New suppliers are selected based on ability to meet current and future needs
  - Potential supplier must meet *all* specified criteria to be selected
  - Items may only be purchased from suppliers on "Approved Supplier List" (ASL)
  - Supplier may be disqualified from ASL for subpar performance
  - Purchase requirements are clearly stipulated in purchase orders

Supplier management and control (contd.)

- Purchased items must pass inspection against purchase requirements prior to use (100% coverage)
- Each purchased item is uniquely tagged and traceable at all times

Technical Processes

- Requirements Definition
- Product Design
- Implementation
- Product Validation (Testing)

Requirements Definition

- Requirements are identified from various sources:
  - Customer requirements (explicit and implicit)
  - Requirements specified internally (explicit and implicit)
  - Statutory requirements
- Requirements are documented in accordance with Product Requirements Specification template
- Product requirements specification is formally reviewed and revised (as needed) prior to approval by key stakeholders
- Formal record of requirements review is created (including list of action items)
- Changes to approved requirements are controlled
- Product requirements are traced during product development and testing

Product Design

- Product architects analyze product requirements specification and create high-level product design specification:
  - Product requirements are allocated to product components
  - Build vs. buy decision is made for each component
  - High-level design is reviewed and revised (as needed) prior to approval
  - Formal record of high-level design review is created
- Product designers and developers create lower level product design specifications
  - Lower level product design is reviewed and revised (as needed) prior to approval
  - Formal record of low-level design review is created
  - Product prototype is developed and verified, if needed
- Changes to approved design specifications are controlled

Implementation

- Product developers develop the product in accordance with product design specifications
- Applicable workmanship standards (identified in the relevant procedures) are used during implementation
- Product is verified to identify nonconformance to design specifications and to identify product defects
- Changes to product design specification are initiated (if necessary)
- Product defects and nonconformances are resolved, and product is reverified
- Records of product verification are maintained
- Verification tools are calibrated and secured
- It is ensured that products and components are uniquely traceable

Product Validation (Testing)

- Product is validated to identify nonconformances against product requirements and to identify product defects
- Product validation plan is created and approved prior to start of validation activities
- Product nonconformances and defects are formally logged and tracked to resolution
- Defective and/or nonconformant product is controlled to preclude delivery to customer or unintended use (unless concession is obtained)
- Product validation is repeated after product nonconformances and defects are resolved
- Records of product validation are maintained; Validation tools are calibrated and secured

Support Processes

- Contract Negotiation and Review
- Product Packaging and Delivery
- Customer Support (including Customer Training)
- Quality Assurance and Quality Control
- Employee Training
- Workspace Facilities and Information Technology Infrastructure

Contract Negotiation and Review

- Negotiated terms and conditions of sale are documented in legal contract
- Contract is reviewed by internal stakeholders prior to approval
- Sample list if items verified during contract review:
  - Organizational ability to meet requirements
  - Clarity, completeness, and unambiguity of requirements
  - Acceptable mechanism for initiating future revisions to the contract
- Records of contract review are maintained

Product Packaging and Delivery

- Procedure has been established to protect product from damage and tampering prior to delivery to customer
- Sample provisions of this procedure:
  - Clearly designated and controlled areas for products ready for delivery
  - Packaged products clearly labeled and uniquely identifiable for traceability
  - Up-to-date inventory of packaged products maintained by shipping department

Customer Support (including Customer Training)

- Departmental procedures have been established for handling of following issues:
  - Customer-reported problems
  - Customer complaints
  - Customer suggestions
  - Request for information (from customers)
- Underlying principles of these procedures:
  - Timely resolution of each customer issue
  - When possible, verify proposed resolution prior to communication to customer
  - Keep customer informed regarding handling of customer issues
  - Maintain records of all customer interaction

Customer Support (including Customer Training) (contd.)

- When possible, collect customer satisfaction data for customer interaction experience
- Examine customer issues for how they can be prevented in the future

■ Procedure for customer training has been established

■ Sample provisions of this procedure:

- Training courses identified after extensive customer needs analysis
- Training tracks established for various types of customers (and end-users)
- Data collected to assess training effectiveness
- Mechanism established to review, prioritize, and implement student suggestions

Quality Assurance and Quality Control

- Quality assurance and control deployed throughout the organization
- Quality is everyone's job, not just the quality department's!
- Business processes have been defined, and documented (when necessary)
- Project quality plans are created for product development projects
- QMS documentation identifies what quality records are required
- Internal quality audit program has been implemented
- Measurements are used for process and product monitoring and improvement

Quality Assurance and Quality Control (contd.)

- Risks are continually identified, and systematically assessed and mitigated
- Supplier performance monitored and controlled
- Culture of preventing problems is encouraged
- As an underlying philosophy, the organization strives for continuous improvement

Employee Training

- Employee competency pertains to employee skills, experience, education, and training
- Employee competency requirements are defined (required and desired competencies)
- Employee competency needs are identified based on the competency requirements
- Competency needs, once identified, are addressed (typically by training)
- Effectiveness of action(s) taken is assessed
- Employee training records are maintained

Workspace Facilities and Information Technology Infrastructure

- Suitable workspace facilities and IT infrastructure are identified for each process
- During process execution, adequacy of workspace facilities and IT infrastructure are monitored
- Workspace facilities and IT infrastructure are maintained
- Mechanism has been established for timely resolution of repair and maintenance requests
- An organization-wide disaster recovery plan has been defined (and tested) for use in the event of a disaster

Questions & Answers

INDEX

D

E

F

G

H

I

J

L

M

N

O

P